SEXUAL STRATEGY

SEXUAL STRATEGY

Tim Halliday

THE UNIVERSITY OF CHICAGO PRESS

The University of Chicago Press, Chicago 60637

This book was designed and produced by
The Rainbird Publishing Group Limited
40 Park Street, London W1Y 4DE

House Editors: Karen Goldie-Morrison Linda Gamlin David Burnie
Design: Rod Josey Ltd
Production: Clare Merryfield

89   88   87   86   85   84   83   82   1   2   3   4   5

Library of Congress Cataloging in Publication Data

Halliday, Tim, 1945–
   Sexual strategy.
   First published in U.K. by Oxford University Press, 1980
   (Survival in the wild)
   Bibliography: p. 152.
   Includes index.
   1. Sexual behavior in animals.   I. Title.
II. Series
QL761.H34   591.56   82-2607
ISBN 0-226-31387-5 (pbk.)   AACR2

Photosetting by SX Composing Limited, Rayleigh, Essex
Illustration origination by Hongkong Graphic Arts Service Centre
Printing and binding by South China Printing Co. Hong Kong

Front cover. A mating pair of golden toads *Bufo periglenes*
from the cloud forest of Costa Rica. The vivid colour of
the male contrasts with the darker colour of the female and is an
example of extreme sexual dimorphism. *M. P. L. Fogden.*

Contents

Foreword

A full understanding of animal species can be acquired only after years of extensive studies in their natural environments. Only in the wild is it possible to discover the evolutionary and adaptive significance of each biological activity. It then becomes apparent that many of the forms, colours and activities of wild animals and plants are adaptive responses to the basic problems of survival: the need to eat, to avoid being eaten, and to mate and reproduce. Each species is beset with a unique set of problems depending on the type of environment in which it lives and on its structure: whether it is in a desert or in a jungle, or whether it is a frog, a tiger or a fly. Each species has evolved its own repertoire of strategies which enable it to survive. A successful individual not only survives but also reproduces to pass its genes on to the next generation. However, only those individuals best adapted to their environment survive, and they transmit the traits which have made survival possible to their offspring, an idea embodied in the phrase 'survival of the fittest'.

It is the aim of this new series *Survival in the Wild* to describe and explain the bewildering diversity of strategies displayed by the living world. Each book selects a biological activity vital to survival and describes the array of physical and behavioural adaptations which have evolved as a result of fierce competition. In an often hostile world, individuals interact with others, as food sources, or potential predators to be avoided, or mates.

The fundamental difference between the sexes, the difference between eggs and sperm, has had a profound effect on the evolution of sexual behaviour, and males and females have evolved very different and often conflicting strategies for reproduction. *Sexual Strategy* examines some of the biological principles that underly the nature of sexual behaviour in animals. The book reflects the fact that in no other aspect of their lives do animals show so much variety as they do in their sexual activities, and describes examples of sexual behaviour in many different kinds of animals. It shows that, beneath this immense variety, there is a set of basic rules that has determined the course of evolution of sexual behaviour.

Tim Halliday has spent several years studying the causal mechanisms underlying courtship behaviour in newts and mate selection in frogs and toads. He is able, therefore, to draw on personal observations and conclusions when discussing the nature of sexual behaviour in animals. In his text, he explains that reproducing is often highly competitive and that, for many creatures, it may involve jeopardizing their own survival: a male cricket who calls to attract a female may also attract a predator that can kill him. But, as Tim Halliday shows, the risks inherent in reproduction arise not only from predators. Other members of the same species can also pose a serious threat. In many animals, there is intense hostility between members of the same sex and between males and females before, during and after mating. A male mantid may be eaten by a female even as he mates with her, and fighting between male red deer and between male elephant seals for the possession of females can lead to severe injuries. It has become necessary, therefore, to question the belief that reproduction is a cooperative venture whose purpose is the perpetuation of the species, and there are many instances where Tim Halliday does put this belief to the test.

In the course of writing this book, Tim Halliday has received help from a number of people either in conversation or in their published work in books and journals. He would like particularly to thank Carl Gerhardt and Steve Arnold for information imparted during useful discussions, Beverley Dugan, Ron Rutowski, Peter Stacey and Diane Williams for allowing him to quote unpublished work, and Nick Davies and Carolyn Halliday for reading the manuscript and for making valuable suggestions.

1 Introduction

This book is about the behaviour that precedes, accompanies and follows the act of mating in animals and man. The word 'strategy' implies that such behaviour is part of a tactical plan towards some specific goal. What is that goal? For a long time biologists, naturalists and the public at large have been content with the explanation that when animals mate their purpose is to 'perpetuate the species'. If this was really their aim, males and females would, by mating, be acting cooperatively towards a common goal. However, the most cursory examination of the sexual behaviour of animals reveals countless examples in which males and females treat each other with great hostility and are clearly not behaving cooperatively. Female spiders and mantids often eat their mates; male sticklebacks are frequently very aggressive towards females; male lions and langur monkeys kill the young of the female members of their social groups before they mate with them. Such apparently antisocial patterns of behaviour hardly seem designed to 'perpetuate the species' and we must seek other kinds of explanation for their evolution. This book examines many examples of the sexual behaviour of animals and seeks to explain their evolution in terms of certain basic biological principles.

The word 'strategy' also has other connotations. To label a sequence of behaviour as a strategy might suggest that it is planned and controlled by deliberate, rational thought, as it would in a human context. When we use the word in relation to the behaviour of animals we are not implying that they use the same kind of thought processes that humans do. Strategy is a convenient label for animal behaviour patterns which are directed towards a clear goal, such as obtaining a mate. It carries no implications about the internal mechanisms that control the behaviour of animals.

Before we can understand the function of sexual behaviour we must examine the purpose of sex itself. Sexual reproduction, in which single cells from two different individuals combine to form new individuals, is but one of a number of mechanisms by which animals reproduce themselves, but it is by far the most widespread. The evolution of sex is still a problem that puzzles biologists and it is discussed in the first chapter of this book. Another basic factor that will be examined is the evolution of gender; that is, the division of individuals into two distinct sexes, males who produce vast numbers of sperms and females who make relatively very few eggs. Differences in gender between individuals lead to differences in behaviour, most importantly in the form of parental care shown by males and females.

Harmony and conflict between mates. The Magellan penguins of both sexes incubate their eggs in a communal site (left). A male Jerusalem cricket (above) has mated with a female but has failed to make his escape and is being devoured by her.

Gender and parental behaviour are two basic ingredients in the evolutionary recipe that determines the form of sexual behaviour. To them we must add a third ingredient, the nature of the environment. The form that mating and its associated behaviour takes is constrained by the many other vital activities that animals must undertake in order that they and their offspring may survive. They must find food, often in competition with other members of their species, and they must avoid death at the hands of predators. Each species is beset by a unique set of such problems. For example, male newts, who court females at the bottom of ponds, have to do so while holding their breath between ascents to the surface to take in air. Many aspects of their complex mating behaviour appear to be adaptations to this basic constraint on their sexual activity.

If the perpetuation of the species is not the goal of sexual strategy, what is? The modern theory of biological evolution sees natural selection as a process in which those individuals who leave the greatest numbers of descendants in subsequent generations pass on their heritable characteristics at the expense of less fecund individuals. The 'aim' of sexual strategy is thus to maximize the number of an individual's descendents. While males and females, as individuals, may share this goal, their sexual strategies for achieving it will be different because of the many differences between them associated with gender. Females typically produce a limited number of eggs and invest a great deal of energy and resources in the production, nourishment and care of each potential offspring. Males, in contrast, produce enormous numbers of sperms and tend to be somewhat profligate with them, seeking to mate with as many females as possible rather than directing a great deal of parental care towards offspring. There are, of course, many exceptions to this pattern of sexual behaviour and some of these will be explored in this book.

In writing a book of this kind one is able to draw upon an enormous number and variety of examples of patterns of animal sexual behaviour and I have been forced to be highly selective. I have tended to use a lot of examples from amphibians, partly because these are animals with which I am very familiar. However, I feel that the inclusion of many amphibians in this book is justified on the grounds that they often provide very good illustrations of particular biological points and also because they have recently become the object of much exciting research throughout the world.

In seeking to find basic biological rules underlying the reproductive behaviour of animals, the question arises, to what extent are such rules applicable to man? Speculation about the evolution of human sexuality is fraught with difficulties and pitfalls, not least because man is so variable in his sexual and social behaviour, suggesting that his habits are largely the product of his culture. In a somewhat cynical view of the sexual ethics of contemporary western man Dorothy Parker wrote in 1944:

> Woman wants monogamy;
> Man delights in novelty,
> Love is woman's moon and sun;
> Man has other forms of fun.
> Woman lives but in her lord;
> Count to ten, and man is bored.
> With this the gist and sum of it,
> What earthly good can come of it?

While this poem expresses sentiments offensive to modern attitudes to the sexual and social role of men and women, it most elegantly expresses the essential conflict of interests that exists between the sexes of the vast majority of species. To what extent this conflict is present in man and, if it is, whether it is the product of biological evolution or of our culture, are questions which will be discussed in the final chapter of this book.

2 Sex

It is characteristic of living organisms, whether they are bacteria, plants or animals, that during their lives they grow, reproduce and die. During an individual's lifetime, reproduction may occur once, several times or more or less continually, depending on the species. There are many ways of reproducing and this book is about only one method, sex.

The mechanisms of reproduction
There are two essential features of sex that differentiate it from other forms of reproduction. First, individuals produce sex cells called gametes by a special sort of cell division called meiosis. Most cells in the body contain pairs of chromosomes; in meiosis these separate. As a result, each gamete contains only half the number of chromosomes of the other cells in the body, and therefore only half the genetic make-up of its parent. Second, gametes from two individuals fuse, producing a new individual whose genetic make-up is made up half and half from each parent. In the great majority of sexually reproducing species there are two types of gamete, female and male. Female gametes are relatively large and immobile, are produced in small numbers and are called ova or eggs. Male gametes are very small and mobile, are produced in vast numbers and are called sperms.

In man, and in a great many species,

individuals are specialized to produce either eggs or sperms and are likewise referred to as female or male. However, a number of species of animals and, to a greater extent, plants are capable of producing both types of gamete within the same individual. Among animals, individuals that can produce both male and female gametes are called hermaphrodites. In some species individuals are hermaphrodites throughout their lives, in others they begin life as one sex and change to the other when they reach a particular age. In the Indo-Pacific reef-living cleaner fish *Labroides dimidiatus* individuals gather in groups of which all but one are female; the male is always the largest fish in the group. If the male is removed from a group the largest of

The two types of cell division. In mitosis (left) the number of chromosomes remains the same but in meiosis (right) the number is halved.

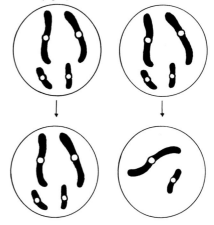

Four sea-slugs, *Aplysia brasiliana,* in a copulatory line. In this mating formation the animal at the back is usually male, the one at the front female, and the animals in between hermaphrodites. Each gives sperm to the animal in front.

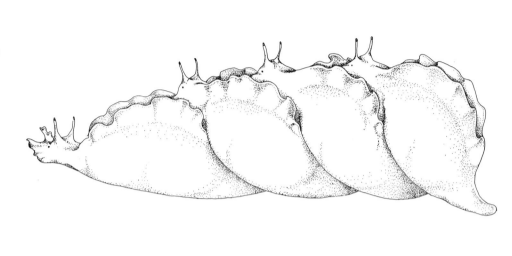

the females quickly changes her gender and becomes a male. The reverse process occurs in another marine fish of the Indian Ocean, the anemone fish *Amphiprion akallopisos,* which lives in groups of which all individuals are male except for the largest who is a female, who was previously a male. In sea-slugs, the situation is even more complicated. In one North American species, *Aplysia brasiliana,* individuals are capable of being male, female or hermaphrodite. Mating may involve only two animals, either one male and one female or two hermaphrodites, or anything up to 20 animals in a mass containing any combination of males, females and hermaphrodites.

It is important to remember that sex, with all its variations, is only one method of reproduction. Many animals and plants show forms of reproduction that involve neither the production of gametes nor the involvement of two individuals. Such forms of reproduction are called asexual and, as with sexual reproduction, they show considerable variety. The simplest form of asexual reproduction, shown by many protozoan organisms, is to divide into two. A number of

animals, like *Hydra,* can reproduce by growing buds which eventually detach themselves as fully formed individuals. Many species produce eggs, as if they were sexual species, except that these develop into adults without being fertilized. This condition is called parthenogenesis, which means 'virgin birth'. There are two kinds of parthenogenesis, differing in the form of cell division that produces the eggs. In aphids, for example, eggs are produced by the normal method of cell division called mitosis, in which there is no halving of the chromosome number. Each egg, therefore, contains a full and unchanged set of parental genes. In bees, as in other sexual species, the eggs are produced by meiosis. However, unlike the eggs of typical sexual species, they are capable of developing into bees without being fertilized. These unfertilized eggs become males or drones, while the fertilized bee eggs become females, *ie* workers or queens.

This picture of the variety of methods of reproduction is made even more complicated by the existence of many species that can reproduce both sexually and

Below Changing sex. Two stacks of slipper limpets *Crepidula fornicata* attached to the same stone, (three on the left and six on the right). The older, larger individuals at the bottom are females; the younger, smaller individuals at the top of the stacks are males.

asexually. Many plants reproduce asexually by sending out suckers or runners and sexually by flowering. *Hydra* is capable of reproducing, not only by budding, but also by producing eggs and sperms. Another variation is shown by hermaphrodite animals or plants that fertilize themselves. Inasmuch as such organisms produce gametes by meiosis they are behaving like sexual species, but in fertilizing themselves they are like asexual species in that there is no combining of genes from different individuals. Many plants are self-fertilizing hermaphrodites, but this mode of reproduction is very rare among animals.

What are we to make of this variety of methods of reproduction? Can we say that one method is better or more efficient than another? This is unlikely because if one mode of reproduction were better than the others we would expect it to have replaced all these others during evolution. The answer seems to be that the variety of reproductive methods is a reflection of the

variety of environments in which animals and plants live. Just as different species of animals vary in their methods of obtaining food or avoiding being eaten by a predator, according to the nature of these features of their environment, so it seems that their methods of reproduction are adapted to environmental factors. There are two basic modes of reproduction, sexual and asexual. The question we must now answer is, under what environmental conditions is each of these methods the more effective? To answer this question we must look more closely at the products of sexual and asexual reproduction. These two processes differ not only in terms of the mechanisms they involve, but also, and most importantly, in the nature of the offspring that they produce.

The end results of any reproductive process is the formation of new individuals. Asexual reproduction produces individuals that are exact replicas of their parents since they have inherited their parent's genes in an unaltered form. By contrast, sexual reproduction involves re-shuffling of each parent's genes during the formation of gametes, followed by the fusion of gametes from different parents. As a result every individual offspring is genetically unique, differing both from its parents and from its brothers and sisters. An essential feature of the theory of natural selection is that during reproduction individuals produce many more offspring than can possibly survive. The resulting struggle for survival can be compared to a lottery, in which each individual offspring represents a ticket whose number is its genetic make-up. In asexual reproduction all the offspring of any one parent carry the same number, that of their parent. In sexual reproduction every offspring carries a different number. Which form of reproduction is more likely to be successful in life's lottery depends on the predictability of the environment.

Any individual who has survived long enough to reach reproductive age must carry a combination of genes that makes it a good survivor, since it has survived while most of its fellows of a similar age have perished. If it lives in a stable environment it would seem that the strategy that would best insure the survival of its offspring is to give them the same good, proven combination of genes. This can be achieved by asexual reproduction. To reproduce sexually would create offspring with combinations of genes that might or might not be suited to the environment. Most will be unsuited and will perish. However, if the environment is unstable, the conditions in which reproducing adults have proved themselves able to survive will not be the same conditions as those that will be faced by their offspring. Thus the survival prospects of genetically identical, asexually produced offspring are likely to be small. By contrast, reproducing sexually creates varied offspring, of which at least a few are likely to be well-suited to the uncertain environment in which they will have to struggle to survive.

The argument that sexual and asexual reproduction are adaptations, respectively, to unstable and stable environments, is supported by many of those species that are capable of both. *Hydra* and *Daphnia* are both small inhabitants of freshwater ponds and pools that are liable to dry up during the course of a year. While the pond is full, and conditions are good, both animals reproduce asexually, *Hydra* by budding and *Daphnia* by parthenogenesis. However, when the pond begins to shrink, conditions deteriorate and both creatures switch to sexual reproduction, producing fertilized eggs. They are contained within drought-resistant coverings so that they are able to survive until the environment once again becomes favourable. A single *Hydra* can produce both eggs and sperm but in *Daphnia* the switch to sexual reproduction involves the production of some exclusively male individuals.

The tiny protected eggs that are produced during sexual reproduction in

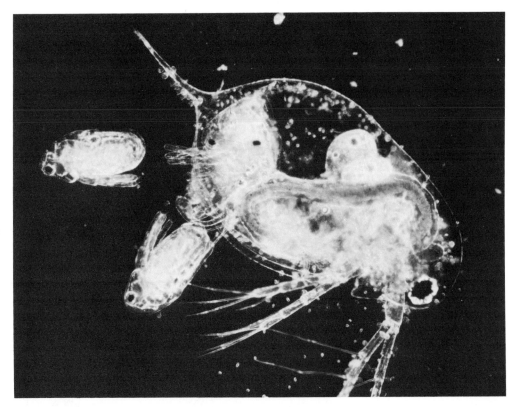

Asexual reproduction. A female *Daphnia* (top) is giving birth to two babies which have developed from unfertilized eggs. The *Hydra* (below) has two buds which will soon become detached and independent. Both these animals reproduce asexually as long as conditions are favourable.

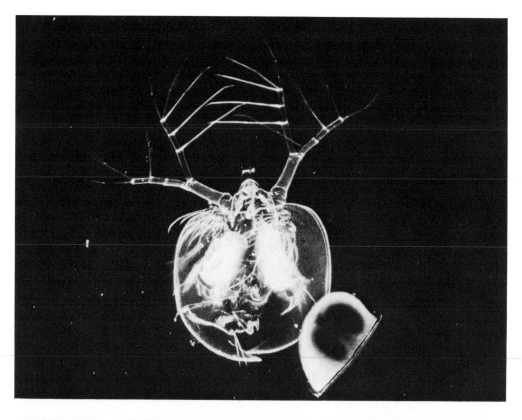

Sexual reproduction as a response to adverse conditions. A female water-flea, as seen from above, with an ephippium or case containing fertilized eggs (top). The *Hydra* (below) has developed three testes and an ovary which contains a large egg. These fertilized eggs will be able to survive conditions of drought.

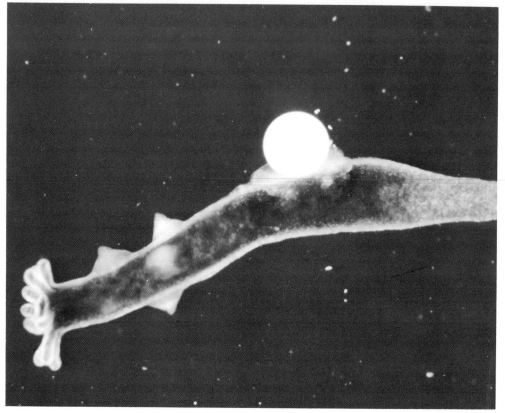

Hydra and *Daphnia* are much better suited than either adults or asexually produced progeny for dispersal to new habitats. Again, using the analogy of a lottery, they all carry different numbers, some of which will be favoured in the different conditions of the new habitat. In most species that have the option of both modes of reproduction, the asexual mode is associated with a sedentary existence, and sex with the dispersal to new and often distant habitats. Many plants, for example, reproduce asexually by sending runners out into the soil immediately surrounding them and sexually by producing seeds that can be carried far and wide by wind or animals. The pond in which a *Daphnia* lives and the patch of soil in which a plant grows are more predictable environments for offspring than the distant habitats whither eggs and seeds will eventually find their way.

Unfortunately, this simple theory that the mode of reproduction of a species is related to the stability of its environment does not fit all the facts. For instance, those plants which we refer to as weeds are those which first appear on newly cleared patches of soil and colonize them successfully. They are highly adapted to disperse themselves and we would therefore expect them to reproduce sexually. Very few of them do. Dandelions, for example, produce their airborne seeds by parthenogenesis.

So far our discussion has assumed that sexual and asexual reproduction are comparable in all but their mechanisms and their products; we have not considered their relative costs. Asexual reproduction is simple and devoid of any intrinsic risks. A budding *Hydra* or a parthenogenetic species does not have to seek a mate as a sexual animal does, often, as we shall see throughout this book, in the face of considerable hazards. However, and more important than costs associated with the act of mating, is the fact that, for females, sexual reproduction may be more costly than asexual reproduction because it involves the

production of sons who may make little or no contribution to parental care. We shall discuss this later in the chapter.

As we have seen, there is one major advantage of sexual over asexual reproduction, at least in unstable environments. The production of genetically diverse progeny increases an individual's chance of having offspring with combinations of genes favourable for survival. This lottery theory is only one of many theories that have been put forward to explain the evolution of sex. We do not have the space to consider all these theories and, in any case, there is still much disagreement among biologists on this subject. But we will consider one other theory which involves a quite different kind of argument. The lottery theory considered what individual organisms might do to produce the greatest possible number of surviving offspring. It suggests that sexual individuals will leave more offspring than asexual individuals when the environment is unstable. This next theory is concerned not with the implications of sex for the individual organism, but for whole species of organisms.

Sexual reproduction continually produces new and unique combinations of genes. In asexual reproduction new gene combinations will arise only as the result of mutations, or changes in the genes, which are extremely rare. A sexually reproducing species is thus much more likely to produce individuals who are well-suited to survive changes in the environment than is an asexual species. As a result, sexually-reproducing species are capable of evolving faster than asexual species. There are examples of asexual species scattered throughout the animal kingdom, though no examples are known among birds or mammals. This scattered distribution, in comparison with the much more widespread occurrence of sex, suggests that species with asexual forms of reproduction have arisen frequently during evolutionary time but have tended to become extinct more quickly than sexual species because they are less able to keep pace with environmental change.

This theory is not incompatible with theories that seek to explain the evolution of sex in terms of how it benefits individuals. What may happen frequently during evolution is that asexual mutants arise within sexual species. Because they are more successful in reproducing themselves than the sexual members of their species, these mutants initially increase in numbers. However, over the course of a number of generations the nature of the environment will gradually change and the progeny of the asexual mutants will be increasingly unsuited to it. The asexual population will produce genetic variants at an extremely slow rate, will thus not be able to adapt to the environmental change, and will become extinct. In contrast, the sexually reproducing population, with its continual production of new gene combinations, will be able to adapt to the changing environment and will survive.

Male and female

The essential feature of sex and the one that seems to have been most important in its evolution is that, because it involves the fusion of gametes produced by meiosis in two parents, it produces offspring that are genetically highly varied. Another virtually universal feature of sexually reproducing organisms is the phenomenon of gender which is the existence of two types of gametes, male and female. These two aspects of sex are quite unrelated to each other. Sexual reproduction would produce just as varied progeny if all the members of a species produced the same type of gamete. Indeed, this condition, called isogamy, is found in some algae and protozoans. Why then, if gender is not essential for sexual reproduction to fulfil its biological function, did males and females evolve and become an almost universal feature of animals and plants?

It is generally assumed that the con-dition that prevailed in the most primitive sexually reproducing organisms was isogamy, like that which persists in algae and protozoans. In this condition individuals cannot be separated into males and females, since all produce identical gametes. It is typical of any sexually reproducing organism that, of a large number of gametes produced, only a tiny proportion will survive to meet other gametes and achieve fertilization. These are the conditions in which natural selection will be very intense, and will strongly favour any adaptation of an individual gamete that increases its chances of surviving to meet another gamete. There are two ways in which gametes could do this: by living longer and by travelling further. To live longer a gamete requires reserves of nourishment. To accommodate these, longer-lived gametes must be larger than other gametes and, consequently, less mobile. To travel further a gamete needs to be highly mobile. Cellular material (cytoplasm) is used not for food storage but to provide a tail that propels the gamete along. All surplus tissue is lost to reduce excess weight, resulting in very small size. Our two types of primitive gamete are well on their way to evolving into the eggs and sperms typical of most living sexual organisms.

An important feature of this theory of gamete evolution is that the two types, egg and sperm, must evolve together. They could not evolve independently. If all gametes within a species had evolved along the line of increasing size and reduced mobility, they would have become increasingly unlikely to meet each other. If all had become tiny and mobile, they might well have been very likely to meet but, in the absence of food reserves, the new individual that they formed at fertilization, which is called the zygote, would soon perish. The evolution of egg-like gametes must have favoured the evolution of sperm-like gametes, and *vice versa*. The relationship between eggs and sperms can be regarded as mutually beneficial because each requires the genetic material contained in the other to form a new individual. However, the relationship is also parasitic; while the sperm contributes only genes to the zygote, the egg provides both genes and the cellular and nutrient basis for its development into a new adult individual. Thus a sperm takes advantage of the material contribution made by an egg to the zygote. This imbalance in the relationship between eggs and sperms forms the basis of a fundamental imbalance that exists in the relationship between male and female adults and which accounts for much of the complexity in the sexual behaviour of animals that we shall examine in this book.

An essential feature of the parasitic relationship between egg and sperm is that eggs are capable of surviving on their own for some time. Females thus potentially have an option that is not open to males; they can become parthenogenetic, their eggs developing without fertilization by a sperm. The cost of reproduction to females in sexual species is, therefore, higher than in asexual species because sexual reproduction involves the production of sons. Whereas all the offspring of a parthenogenetic female will be daughters, only half of the offspring of the sexual females will be daughters, the other half being sons. In terms of reproducing herself, a parthenogenetic female is twice as successful as any sexual female. Any advantage that sex might have over asexual reproduction must be sufficient to offset this intrinsic cost of sex. We shall see in later chapters that in many species, notably in birds, the males do help with the rearing of young, and there may be no cost of sex if the pair can raise twice as many young together as the female could have alone.

Male and female animals do not differ only in the kinds of gametes they make, but also in the number they each produce. Females typically produce only a few eggs whereas males generate many millions of sperms. To make gametes requires resources in the form of energy and nutrients. Females allocate a relatively large amount of these resources into a few, large eggs. Males allocate only a tiny amount to each of their innumerable sperms. This difference between the sexes in their pattern of resource allocation to their gametes is basic to our understanding of the evolution of the relationship between the two sexes.

Reproductive effort

Throughout their lives animals acquire the energy and nutrients essential for life, growth and reproduction by feeding. These resources are used up during an animal's day to day activities which we may call its effort. There are two kinds of effort, somatic and reproductive. Somatic effort is that which maintains the survival and growth of the individual, reproductive effort is that which is put towards reproduction. Reproductive effort may be directed towards three different aspects of reproduction, the act of mating and its preliminaries, parental care of the developing young and, in some species, care of the young of other, usually related individuals. Reproductive effort has been described by some biologists as reproductive investment. When reproducing, animals invest time, energy and resources and gain a return on this investment which is measured in numbers of surviving offspring. Males and females apportion their reproductive effort or investment in fundamentally different ways. Females invest heavily in each of a few potential offspring, males invest very little in each of many.

The most important difference between the reproductive effort of the sexes is the amount of effort that goes into parental care. Females make eggs that are large and which contain the basic resources required for the first stages of the development of the zygote into an adult. They have, therefore, made a substantial parental investment in their potential offspring even before fertilization occurs. Males contribute only genes to the zygote; they have

made no initial parental investment before fertilization. In the great majority of animals we find that this imbalance between the sexes in terms of parental investment has been accentuated during the stages of development of the offspring that follow fertilization. The female may, for example, equip the egg with an additional food supply in the form of yolk. A bird's egg represents a huge allocation of nutrients, by the female, to the provision of food and protection of the developing embryo from its conception to hatching. In mammals, the egg is retained within the mother's body during the first part of its development and is fed by her through the placenta. Development within the mother, called viviparity, occurs also in some fish, amphibians and reptiles. The female mammal is also specialized to nourish her young after their birth by being able to secrete milk.

Males show none of these specializations for caring for young. As a general rule females direct much more of their reproductive effort into parental care than do males. In many species it is the female alone who carries out such parental tasks as the incubation of eggs and the feeding of young. There are, however, many exceptions to this general rule. In some species, such as sticklebacks and the midwife toad *Alytes obstetricans*, it is the male who tends and protects the eggs. A feature common to these animals is that the eggs are fertilized outside the female's body and this may partly account for this reversal of what we tend to think of as the 'normal' sex roles. Since a female stickleback or toad lays her eggs before the male fertilizes them, he is essentially left 'holding the babies'. That is to say, he can either look after them or leave them to face the vagaries of the environment alone. In animals like birds and mammals, in which fertilization occurs inside the female's body, the situation is reversed. She is carrying the fertilized eggs and so is typically the partner who devotes most care to them. However, the timing of fertilization in relation to

Maternal care. A female giraffe provides food for her calf (left). A female Nile crocodile protects her newly hatched babies by carrying them to water within the safety of her mouth.

Paternal care. A male midwife toad with his eggs wrapped round his hindlegs. The male glass frog *Centrolenella columbiphyllum* guards his eggs in vegetation above forest streams, into which the tadpoles drop when they have hatched.

egg-laying is only one factor in determining which sex is the primary provider and protector of the young and we shall examine a number of other factors in later chapters. In many birds both partners care for the young. In pigeons, for example, the chicks or squabs are fed on a milk-like secretion that is produced in the crops of both parents. As in other aspects of sexual reproduction, there is enormous variety in the extent to which males and females share in the parental care of their offspring.

While males generally make less parental investment than females, this does not necessarily mean that they expend less reproductive effort overall. Parental care is only one component of

Two forms of parental care. Unlike many fish, the dogfish does not leave its eggs entirely unprotected. Each one is supplied with ample yolk and is surrounded in a 'mermaid's purse' (left). A wood pigeon *Columba palumbus* (above) feeds its young on crop milk. This method of feeding is unique to the pigeon family and is unusual in that both male and female secrete this milk.

reproductive effort; in terms of another component, mating effort, males generally make a much greater investment than females. This arises from the fact that a female's reproductive potential is much smaller than that of a male because she produces far fewer gametes. Females frequently only produce sufficient eggs to mate and reproduce once during a breeding season, and often only once during their lifetime. Males have the potential to mate with a large number of females. Thus, while there is usually an equal number of males and females in a population, potential mating opportunities are not equal for the two

sexes. For a female, with her limited mating opportunities, the optimum strategy is to mate with a male whose genes will contribute most to the survival and viability of her offspring, in other words, the fittest male she can find. For a male the optimum strategy is to mate with as many females as possible. There is thus a basic conflict of interests between males and females because each sex is pursuing a different strategy. In effect, females represent a limited resource for which males must compete among themselves. This competition among males imposes a powerful selection pressure that favours those males who make the most effective mating effort. This selection pressure is called sexual selection.

Sexual selection

Sexual selection is the evolutionary process that favours adaptations that increase the mating success of individuals. As we have seen, it is usually much more intense among males than among females. Charles Darwin divided sexual selection into two types. In the first, males compete aggressively among themselves for sexual access to females. This is called intrasexual selection and it favours the evolution of male characters, such as large size and special weapons like horns, that increase male fighting ability. In the second type, called intersexual selection, males compete to attract females. This favours the evolution of elaborate male displays, colour patterns and special structures, like crests and plumes, that enhance a male's attractiveness to females. As we shall see in the next chapter, which of these two types of selection pressure operates within a given species depends largely on the nature of the species' environment. The result of sexual selection is that males expend considerable reproductive effort before mating, either in aggressive interactions with each other or in elaborate displays to females.

We have seen that sexual selection arises because females represent a limited resource for which males must compete. A result of this competition is that, while all the females in a population usually mate, it is frequently the case that only some of the males do. Some males may achieve several matings, others none. Only those males who are successful in mating competition pass on their genes to succeeding generations. As a result, sexual selection is a very intense form of natural selection and leads to the very rapid evolution of male characters that enhance mating success. Another factor which further intensifies sexual selection is the sex ratio. The more males there are relative to the number of females, the greater the competition between males and therefore the greater the variation between males in their mating success.

In the majority of species the sex ratio is one to one, or very close to it. We tend to take this for granted as a fact of life but, like any other biological character, the sex ratio is something that requires an evolutionary explanation. Why should there be equal numbers of each sex? Why should there not be a sex ratio biased towards one of the sexes? We can illustrate this problem by considering a species in which there is very intense aggressive competition among males. In the northern elephant seal *Mirounga angustirostris* the males engage in severe and bloody fights that result in the formation of a dominance hierarchy in which only the most dominant males gain sexual access to females. As a result, only a tiny proportion of males, about 10 per cent, fertilize virtually all the females in the breeding population. Put another way, only one male in nine leaves any offspring. It seems therefore to be maladaptive to produce equal numbers of sons and daughters, since nine out of ten sons seem to represent wasted reproductive effort. Surely it would be a better strategy to produce sons and daughters in the ratio nine to one? While this seems a more economical form of reproduction from the point of view of the species, it would not be the optimal strategy for individuals to pursue and it would not therefore evolve by natural selection. Every daughter that a pair of elephant seals produces has the potential to produce one pup in each of her breeding seasons, but each son may produce nine pups if he is successful or none if he is not. It is advantageous to individual elephant seals to produce several sons because, by doing so, they increase their chances of producing a successful one.

Large antlers are not necessarily an extreme result of sexual selection. Reindeer and caribou (or New World reindeer) *Rangifer tarandus* are the only members of the deer family in which both sexes have antlers.

His success will cancel out the failure of the unsuccessful sons. Because both sons and daughters, on average, are equally successful, the optimum strategy is to produce them in equal numbers.

In this chapter we have seen that sex is but one of a number of evolutionary adaptations for achieving reproduction. Its widespread prevalence among animals is probably related to the fact that it produces genetically varied offspring. We have seen that sex is almost invariably associated with differences in gender and that eggs and sperms represent fundamentally different, though complementary adaptations of gametes that enhance their chances of achieving fertilization. Because male and female gametes differ not only in form but in the numbers in which they are produced, male and female parents are predisposed to allocate their reproductive effort in different ways. Females typically make a much greater parental investment in each potential offspring than males do. Males generally make a much greater mating effort than females do, and variation in male mating success creates the conditions under which sexual selection acts upon the form of male mating effort. What is important to our understanding of the sexual behaviour of animals is not so much that males and females expend different amounts of each kind of reproductive effort, for example in parental care, but that they allocate their total reproductive effort in different proportions to the different aspects of the reproductive process. These discrepancies between the sexes are different in degree from one species to another. In some species, we find males and females forming monogamous pairs for life and sharing in many aspects of parental care. In others we find males defending harems of several females by aggression with other males and taking little or no part in parental care. In other species, we find females each controlling a number of males, and then it is the males who perform all the parental duties. In Chapter 3 we shall examine the biological bases of these differences between species.

3 Mating systems

Although the act of mating is the most important single event in the process of reproduction, we would gain a very limited understanding of the evolution of animal sexual behaviour if we only considered those interactions that occur between males and females during mating. We saw in the previous chapter that sex is typically the cause of intense competition between animals. It is therefore very important that we consider not only the relationships that animals have with their mates but also those that they have with animals with whom they compete. For many animals, especially the females, the reproductive effort that is expended during mating is insignificant compared with that which goes into parental care. Behaviour associated with mating has to be seen in the wider context of the extent to which males and females each care for their young. For many animals reproduction is only one of the reasons why they form social relationships with other members of their species. For example, they may feed more efficiently or be less likely to be caught by predators if they live in groups. In this chapter we look at the broad context within which mating occurs, taking into account the extent to which animals compete for mates, care for their young and associate with other animals in ways not related to reproduction. As in other aspects of biology, there is enormous diversity in the patterns of sexual, ag-

gressive, parental and social behaviour shown by animals and our task will be to uncover some of the basic biological 'rules' that underly this diversity.

The pattern of sexual and social relationships that forms the context within which mating occurs is called a mating system. Three basic types of mating system have been recognized: monogamy, polygamy and promiscuity. Monogamy is a system in which both males and females mate only with one partner during a breeding season. The bond between a male and female may last only long enough for mating to take place, but may be stable for the duration of at least one breeding cycle, or, in some species, may endure for life. In polygamy an individual of one sex mates with several members of the opposite sex. When one male mates with several females the system is called polygyny; when one female mates with several males it is called polyandry. Promiscuity refers to species in which no durable mating relationships are formed; members of both sexes may mate with several partners and matings occur more or less randomly. It is now a largely disused term. Apart from its human connotations, which are not appropriate when applied to other species, it is an unsatisfactory term to use in relation to animals because, where supposedly random mating systems have been analysed carefully, it has often been found that there are in fact clear patterns in terms of which individuals mate with each other.

Determinants of mating systems

Attempts to classify mating systems according to their structure have not been very successful in helping us to understand the factors that have lead to their evolution. Many species have mating systems that do not fit readily into any of the three simple categories listed above. There are species which fit one category at one time or locality but not at another, and there are some species in which, for example, some individuals are monogamous and others are polygamous. A more useful approach is to analyse mating systems in terms of the biological factors that determine their structure and to classify them on the basis of these factors. We shall first discuss a number of ecological and social factors that influence mating systems and then examine some examples of different mating systems in detail.

To sustain the effort that they put into reproduction, especially that directed towards parental care, animals require food, cover and other resources from their environment. The richness of ecological resources essential to reproduction is one of the principle determinants of mating system structure. If food is abundant and easily obtained it may be possible for one parent to rear the young successfully alone. In a rich environment, therefore, it may be adaptive for members of one sex, usually males, to mate and then take little or no part in the care of the resulting young but to devote their efforts to obtaining further matings. Thus rich environments can provide the conditions in which polygyny may be a more adaptive strategy for individual males to pursue than monogamy. Where food is scarce and difficult to find, offspring may only be successfully reared if both parents participate in feeding them. In such conditions it will not be adaptive for a male to leave a female looking after their young and to seek further matings because by doing so he will leave fewer offspring than if he had stayed with her. A rich food supply does not inevitably lead to the evolution of a polygynous mating system; other ecological factors are also important. For example, if predation is a serious threat to breeding animals and their young, it may be more advantageous for a male to share in the protection of the young than to leave the female on her own. Thus severe predation can favour monogamy, quite independently of food supply.

As well as the overall abundance of resources necessary for reproduction, an important ecological determinant of mating systems is the spatial distribution of resources in the environment. If such requirements as food, water, hiding places and nest sites are distributed in patches, animals must converge on those patches to exploit them. In species in which there is already a potential for polygyny, such aggregations provide an opportunity for individual males to control sexual access to females. It is much easier for a male to gain an advantage over other males, by competing aggressively with them, if females are gathered together in a restricted area. Thus a patchy distribution of resources tends to be associated with polygynous mating systems in which a male's success is dependent on his ability to compete aggressively with other males. Males may compete directly for females, rounding them up and defending them against rivals. Alternatively, they may compete indirectly by defending part or all of the resource patch on which the females are converging. In some species, males defend territories in areas where resources are abundant, allowing females to enter but keeping other males out.

A patchy distribution of resources is not the only ecological factor that can cause animals to form groups and thus create conditions favourable for polygamy. For many animals, social behaviour is adaptive either because it

Opposite Division of labour in a monogamous pair. A male yellow-billed hornbill *Tockus flavirostris* brings food to his mate who is walled up inside the tree while she incubates their eggs.

affords some protection against predators or because it increases feeding efficiency. Stealthy predators and camouflaged prey are more likely to be detected by several pairs of eyes than by one. Animals living in a group may join forces in warding off a predator. If females form themselves into groups for such reasons they make it easier for individual males to establish control over them.

There is, however, a limit to the number of females that any one male can monopolize. The larger the group of females or the territory that a male attempts to defend against his rivals, the greater the effort he has to expend in aggression. This is important in relation to another factor that can contribute to the nature of mating systems. If females are highly synchronized in coming into reproductive condition, there will be many of them available to males at one time and individual males will only be able to monopolize some of them. Thus a large proportion of the males in a breeding population will obtain matings. Conversely, if individual females are not synchronized but come into condition at different times, only a few will be available to mate with males at any one time. In these conditions individual males may, over the course of a long season, gain a very high proportion of matings by controlling access to the limited number of females available on any one day. Thus polygynous mating systems are more likely to evolve in species where females do not have synchronized reproductive cycles.

A quite different factor that can be an important determinant of mating system structure is the extent to which the young are capable of looking after themselves. Among birds there are two kinds of young: the precocial chicks typical of ground-nesting species, which are capable of running about and feeding themselves soon after hatching, and the altricial young of most tree-nesting species that hatch when they are still blind, naked and largely helpless. Since precocial young make fewer demands on their parents, it may be easier for one parent to rear them alone. As a result, polygamy, in which parental care is not equally shared between the parents, is rather more common among precocial than among altricial species.

Monogamy

The simplest mating system is monogamy because it involves the smallest number of animals. While it is the commonest mating system among birds, being found in 90 per cent of species, monogamy is very rare among other vertebrates. The prevalence of monogamy among birds is probably due to the fact that they produce young which, both as eggs and as chicks, require a very large expenditure of parental effort if they are to survive. Eggs have to be constantly incubated and defended against predators, and chicks require warmth, protection and, in many species, continual feeding. Only in the most favourable environments, where it is warm, where predators are few and where there is abundant food, is one parent able to rear the young alone. For most bird species it seems that the combined efforts of both parents are required to ensure the successful fledging of the young. Once a clutch of eggs is laid, any attempt by either partner to seek additional matings would detract from its ability to rear that clutch and would thus jeopardize the eggs' chances of surviving to fledging.

In some monogamous birds, parental duties are shared by the two parents. In herring gulls *Larus argentatus*, for example, both the male and the female incubate the eggs, defend the nest and go out to collect food for the chicks. While one is gathering food, the other is incubating and defending, and the two parents switch roles frequently. In other birds the different parental roles may be performed by different parents. In some of the African hornbills, the nest is built in a deep hollow within a tree. The female walls herself in with mud brought to her by the male, leaving a hole just large enough for him to pass food in to her. She thus incubates the eggs alone while he provides the food for her and the chicks. The female does not break her way out of the nest until the young are so large that there is no longer room for her. She then joins the male in feeding the chicks, who are again walled up in the nest until they are fully fledged.

Some birds maintain stable pair-bonds over several years, and sometimes for life. In the kittiwake *Rissa tridactyla*, called the black-legged kittiwake in North America, it has been shown that this fidelity increases the reproductive success of a pair. Individuals who have bred before but who have to form new pairs because the original mate died, succeed in fledging fewer young than those individuals who do not change their partners. For kittiwakes, as for other long-lived birds, experience of breeding increases reproductive success and this effect is enhanced if they are thoroughly familiar with their mates.

In monogamous species most individuals in a breeding population have an opportunity to mate and there is very little competition among males for access to females. As a result there will not be strong sexual selection among males. Consequently males and females in monogamous species tend to look alike (they are not sexually dimorphic). In most gull species, for example, it is very difficult to tell one sex from the other. The same is true of such monogamous species as the European robin *Erithacus rubecula* and the great crested grebe *Podiceps cristatus*. However, there are monogamous species that show sexual dimorphism, perhaps as an adaptation that aids species recognition, though in some instances sexual dimorphism may have nothing to do with sexual behaviour. In a number of monogamous tree-nesting ducks, the female is considerably smaller than the male and this is thought to be an adaptation that enables her to enter small holes in

trees where she can incubate the eggs safe from the attacks of predators. The huia *Heteralocha acutirostris* is an extinct species from New Zealand in which the male had a short stout beak with which he hammered into wood in search of insect prey, whereas the female had a long, curved beak adapted for probing into crevices. These differences in beak shape and feeding technique meant that a male and a female did not compete with each other for the same food sources. Reduction of competitition between the sexes for food is also probably the adaptive value of the dimorphism in size that occurs in many monogamous birds of prey. Females are generally larger than males and tend to catch slightly larger prey.

The evolution of monogamy is not only determined by whether or not it is in the interests of the male to remain with a female after mating. It will virtually always benefit a female if a male stays with her and helps her to feed and protect the young because she will be able to rear a larger clutch than if she were on her own. For this reason females often synchronize their breeding activities. This is particularly true of species such as herring gulls that nest in dense breeding colonies, where the opportunities for males to find additional mates are greater than they are in species which nest widely spaced apart. Synchronized breeding in colonial birds is probably brought about by the sexually stimulating effect of a very large number of birds all displaying to their mates in very close proximity to each other. It has been suggested that the adaptive value of breeding synchrony is that it reduces the risk of each individual's eggs being taken by predators. Predators will only be able to eat a certain number of eggs each day. If all the birds in a colony lay their eggs on the same day, this increases the number of eggs available to the predator on that day but reduces the probability that any one clutch of eggs will be taken. It is known that black-headed gulls *Larus ridibundus*

Overleaf Sexual and social harmony. Kittiwakes form pairs for life and males and females share all parental duties. They look identical, feed one another, and nest in dense colonies on cliffs where each pair is very tolerant of its neighbours.

Sexual dimorphism is not always the result of sexual selection. In the extinct huia of New Zealand, the male had a short, stout beak and found insects by hammering into branches while the female used her long, curved beak to probe into crevices. As a result, they avoided competition for the same food.

that lay either a few days before or a few days after the day of peak laying are more likely to lose their eggs than those who lay on the peak day. Another reason why breeding synchrony may have evolved is that, if a female mates on the same day as most other females, this reduces the chances that a male will be able to find a female who has not yet mated. A female who mates early in the season runs the risk of losing her male because he has ample opportunity to find another, as yet unmated female. Thus breeding synchrony in females may be an adaptation that increases the likelihood that males will be monogamous.

Polygyny

As we have seen, polygyny becomes a more adaptive strategy for males than monogamy when resources are patchily distributed. Under such conditions males have the potential to monopolize several females by defending and controlling access to essential resources. A striking example of this kind of system, called resource-based polygyny, is provided by the orange-rumped honeyguide *Indica-*

tor xanthonotus in Nepal. Beeswax is an essential component of the diet of both sexes in many honeyguide species, and the orange-rumped honeyguide obtains it by raiding the nests of a giant honeybee. Bee nests are rare and are found only on cliff faces; they are thus a scarce and patchily distributed resource. Males take no part in the feeding or protection of their young but centre their activities around bee nests, where a small proportion of males establish territories. Other males are excluded from these territories but females and immature birds are allowed to enter throughout the year. In the breeding season the territorial males court and copulate with the females that come to their territories for wax. Non-territorial males also display to females but are rejected by them. Thus only the territorial males achieve any matings and in a study of this species one territorial male was observed to mate with 18 females, while non-territorial males were never seen mating.

The orange-rumped honeyguide represents an extreme example of resource-based polygyny in that it is based on one very specific ecological resource and because males take no part in parental care. In other species, males control a wider range of resources and also contribute to the defence or feeding of their young. In the long-billed marsh wren of eastern North America, *Troglodytes palustris*, males establish territories before the breeding season and within them build several nests, far more in fact than will eventually be used for breeding. There is much variation in territory size and males with large territories attract more females than those with small ones. There is also variation in territory quality, in terms of the abundance of food. It is thought that females are attracted to individual male territories partly by the number of nests that the resident male has built in them. As a result male mating success is correlated with nest number. Thus bigamous males build, on average, about 25 nests, monogamous males about 22 and males who do not succeed in attracting any females average only 17 nests. It is thought that the number of nests a male builds is a reliable indicator of the abundance of food within his territory, because males for whom food is easy to find have more time for building nests. Thus females who pair with males with several nests are gaining access to a territory with a rich food supply on which they can feed their young. In defending their territories males are contributing to the care of their young by ensuring them an adequate food supply. This example illustrates two important points. First, although the mating system is largely controlled by male territoriality, the female's behaviour is also important because she exercises some choice over which males she will mate with. Second, not all males in the population are polygynous; some are monogamous and others obtain no mates, showing that the distribution of matings within a population can be very variable.

Food is not the only ecological factor on which resource-based polygyny may depend. In an American bird, the dickcissel *Spiza americanus* the number of females that a male attracts is related to the density of the vegetation within his territory. Males defending territories where vegetation is dense attract more females than those with sparsely vegetated territories. Dense vegetation provides more secluded nest sites than sparse vegetation and is thus more attractive to females. Male three-spined sticklebacks *Gasterosteus aculeatus* who defend large territories attract more females to their nests than those with small territories. Here the important factor is predation by other male sticklebacks who tend to eat the eggs and young of their neighbours. Males who hold large territories do so because they are more aggressive than other males and are thus more likely to repel the attacks of other males. Females, who take no part in the care of their young, can thus increase the survival chances of their eggs by mating with males who hold large territories.

A different type of polygyny, which is not primarily determined by male control of resources, involves the direct control of females by males and is called female-defence polygyny. The group of females that a male defends and mates with is often called a harem. The formation of harems is facilitated if females are predisposed to gather in groups for some reason not necessarily related to mating. Impala *Aepyceros melampus* are browsers of African wooded grassland. Females tend to form groups as an adaptation against such predators as lions and cheetah. These groups do not move around at random but spend the greater part of their time in areas where food is abundant and where there is also good cover. Male impala are territorial and those males whose territories contain good grazing and cover are more likely to have groups of females passing through them than those with poorer territories. However, a male's mating success is not determined only by the

Overleaf A group of female impala has come into the male's territory to look for food. This has given the male the opportunity to acquire a harem which may number from 10 to 100 females. They may stay for the whole mating season if the food supply is good.

A red deer stag roars
to proclaim his
ownership of a
harem of hinds.

quality of his territory; he continually has to repel the attacks of other males who do not hold a harem. A similar situation occurs in red deer *Cervus elaphus* in Europe. A stag competes initially for the control of good grazing areas to which females are attracted and then, once a number of hinds have gathered, for the defence of his harem. He also continually chivvies the hinds, chasing after any of them that wander away and herding them back into his territory. Fights between red deer stags can become extremely violent and prolonged and may result in serious injury. Furthermore, a fighting stag may lose some or all of his hinds to other stags who, taking advantage of his pre-occupation during a fight, move in and chase hinds back to their own territories.

Female seals have to leave the water once a year, hauling themselves out on to a beach where they give birth to their pups. Suitable beaches are few and far between and consequently large numbers of females are attracted to confined areas. Since mating occurs shortly after birth at these same sites, they become the focus of intense competition among males. In the elephant seal, *Mirounga angustirostris*, fights between males are extremely violent affairs and the sea around a breeding beach often becomes red with their blood. Some of the newly born pups are crushed to death in these skirmishes. Only the most powerful males are successful and these become 'harem masters'. Less than a third of all the males in a breeding population mate during a season, and less than 10 per cent of the males fertilize nearly 90 per cent of the females. High-ranking males attack subordinates who attempt to mate with females and will even interrupt their own mating activities to do so. Fighting exacts a severe toll on males, despite their enormous size and strength. Some bulls die immediately after a year in which they have been harem masters, apparently weakened by their exertions. Males very rarely achieve the status of harem master for

Overleaf A female's
view of a sage
grouse lek. A number
of males are
displaying. Their
booming can be
heard from a great
distance.

more than three consecutive years, while females may breed each year for up to 10 years. Some males never acquire a harem and die without reproducing.

In some polygynous species males defend neither a resource nor a group of females. Instead, the distribution of matings among males is determined by dominance relationships that males establish among themselves. This kind of system, called male-dominance polygyny, is best typified by those species that form leks. While leks occur among insects, fish and mammals, they have been studied in most detail in birds such as the ruff *Philomachus pugnax* in Europe and in several species of grouse, including the American sage grouse *Centrocercus urophasianus*. A lek is a confined area in which a number of males compete with each other for the possession of very small territories or mating stations. These serve only as mating sites; they contain no ecological resources of use to females. An important factor in the evolution of lek behaviour is that females are not synchronized in their breeding activity. While all the males stay at the lek throughout the breeding season, each female comes to it when she is ready to mate and stays only as long as it takes her to select a male and to mate with him. She then leaves and begins the task of rearing her young alone. Thus, at any one time during the season, there are many more males than females at a lek and, as a result, competition between males to attract the visiting females is intense. In several lekking species, the males that are most successful are those that hold territories at the centre of the lek. These are especially attractive to females and are the object of the fiercest competition between males.

In any polygynous mating system there is bound to be considerable variation in the mating success of males; some mate with several females, others with none. As a result, sexual selection favouring any male charac-

Left A male sage grouse increases his visual appeal by contrasting both colour and shape. The brilliant white of his inflated chest contrasts with his dark and pointed tail feathers.

Below Explosive breeding. Several pairs of frogs converge and lay their eggs together. One advantage of such highly gregarious spawning is that the temperature in the middle of a large mass of spawn is appreciably higher than that of the surrounding water. As a result, the eggs hatch earlier.

teristic that increases mating success, is a powerful force in polygynous species. It is therefore not surprising that it is among such species that we find some of the most extreme examples of sexual dimorphism. In resource and female-defence polygyny the male attributes that are advantageous are those that increase strength and fighting ability: the males are generally much larger than females and often possess special weapons such as horns, antlers or elongated canine teeth. In lek species, though males do compete for territories, male mating success is largely dependent on being attractive to females. Males in lek species often have elaborate and brightly coloured plumage, as exemplified by such birds as the ruff and the sage grouse.

Mating systems that involve defence, be it of resources, females or tiny mating sites, are generally associated with a reasonably long breeding season, of several weeks or months. Aggres-

sive behaviour can take up a lot of time, and relationships of dominance and subordinance between individuals take some time to become fully established. In species which, for climatic or ecological reasons, have very short breeding seasons, there may simply not be time for anything other than mating activity. Species which have very short breeding seasons and which congregate to breed in large numbers are called explosive breeders. The European common frog *Rana temporaria* provides a good example of such a species; frogs may have a breeding season that lasts only two days. Very sensitive to changes in the weather, they suddenly appear in large numbers in spring following the onset of mild, damp weather. They converge on ponds or ditches where dense, writhing masses of mating and spawning frogs are formed. While there is much struggling between males to gain mating positions on the backs of females, there are very nearly as many

males as there are females in a mating group. This parity in the sex ratio, combined with the brevity of the mating season, means that there is no possibility that individual males can in any way monopolize several females. However, mating in frogs is not random. Males and females sort themselves out by size, large males tending to spawn with large females, small males with small females. How this assortative mating comes about is not entirely clear, but it may be that a male who is matched for size with the female that he is grasping has a firmer grip than one who is smaller or larger than the female, and is therefore less likely to get dislodged in the struggles that occur between males who are clasping females and those who are not. Explosive breeding does not fit very easily into any of the categories of mating system we have examined, but because males often fight and may in some species mate with more than one female, it is usually regarded as a form of male-dominance polygyny.

While the essential feature of polygynous mating systems is that some males mate with several females, in most cases individual females only mate with one male. Certain ground-nesting birds, however, show a mating system in which both sexes may mate with more than one partner during a breeding season. This mating system is called rapid multiple-clutch polygamy and is exemplified by the red-legged partridge *Alectoris rufa* of Western Europe. Pairs are formed on territories established by the males. After mating, a female lays two clutches of eggs in the two nest-scrapes on the ground made by the male. The male incubates the first clutch, the female the second. If all goes well, that is if neither nest is raided by a predator, both clutches will be successfully hatched. The young are precocial and are soon independent and, if conditions are still suitable for breeding, the pair may mate again and produce two more clutches. Though individuals

Opposite The swampland habitat of the American jacana is a hazardous place to make a nest, and the high rate of egg losses to predators has favoured the ability in females to lay many eggs. Males fulfil all other parental duties. Jacanas are polyandrous, the female maintains a large territory which contains several nests each defended by a different male.

tend to stay with their original partners, they may split up and produce their second clutches with new partners. This is more likely to occur if one of the first clutches is lost. If the male loses his clutch he may mate with a second female, while the first incubates her surviving clutch. Likewise she may go to the territory of another male if she loses her first clutch. This system is thought to be an adaptation to the rather unpredictable environment that birds like partridges live in and to the heavy predation that they tend to suffer as a result of nesting on the ground. The ability of females to lay several clutches of eggs in a year is highly advantageous in a year when conditions for breeding happen to be good and in the event of a clutch being destroyed. Males, by taking a full share in the care of the young, enable females to expend more of their reproductive effort in egg production than they could if they had to care for the young alone.

Polyandry

We turn finally to the rarest but in many ways the most interesting type of mating system, polyandry, in which one female mates with and largely controls the breeding activity of several males. The fascination of polyandry is that it involves a reversal of what we tend to regard as the normal roles of the two sexes. One of the best studies of polyandry is that of the American jacana *Jacana spinosa* in Costa Rica. Jacanas live in lily-covered lakes and lagoons where their enormously elongated toes enable them to walk on the floating vegetation. Females defend large territories which are subdivided into smaller territories, each defended by a male and containing a floating nest. Females move around their territories, laying in each nest a clutch of eggs which is then incubated entirely by the male defending that nest. The female is thus emancipated from all parental duties, devoting her time to mating, egg-laying and defending her large territory against other females. She is dominant over her males, being larger than them, and will break up any fights that develop between them.

What is the evolutionary basis of this remarkable reversal of sex roles? As with other kinds of mating system, the answer to this question is related to the kind of habitat in which birds like the jacana live. Suitable breeding sites are rare and the nests are exposed to heavy predation. Such conditions impose a powerful selection pressure that favours the ability of females to lay many eggs, both to exploit those suitable breeding sites that do exist and to replace the eggs that are lost. A female's ability to do this is the greater the less she has to direct reproductive effort into parental care. Furthermore, the capacity to produce many eggs may be associated with the large body size of females. If females are larger than males, it also enables them to assume a dominant role over males. Conversely, because males are preoccupied with parental care their capacity to expend effort in obtaining matings is reduced. The disparity in size between males and females has almost certainly been further accentuated by intrasexual selection. Females have to compete aggressively with one another for a limited number of breeding territories and, just as competition between males in polygynous species favours the evolution of large size in males, so it favours large female size in polyandrous species. Female jacanas are from 50 to 75 per cent heavier than males. In the northern or red-necked phalarope *Phalaropus lobatus* there is no territorial behaviour but an explosive breeding system that is rather like a lek, except that it is the females who compete to attract the males. In this species, sexual selection has favoured the evolution of brighter plumage in females than in males.

Anomalous mating systems

The various species we have discussed in this chapter have been chosen because they each illustrate a particular type of mating system. However, there are species whose mating systems do not fit into any of the categories we have described. This is the case in many species of ducks such as the mallard *Anas platyrhynchos*. Pair formation in the mallard takes place during late autumn at communal display sites. Males gather on a river or lake and establish a dominance hierarchy among themselves. The arrival of females among the groups of males provides the stimulus for vigorous collective courtship by drakes who perform a variety of different displays which they direct towards the females. Males fight and chase each other in an attempt to get into positions from which their displays will be most visible to females. Subordinate males tend to be forced to display from positions which are at the edge of a female's field of view and are therefore less likely to attract her attention. Since the females are continually swimming about on the water, the optimum display positions are constantly moving. Females respond selectively to the displays of certain males, eventually copulating with them. Female preference is based partly on a male's status, dominant males being preferred to subordinate ones, and partly on the brilliance of his plumage, the most brightly coloured males being favoured. Thus a mallard mating group is rather like a lek, except that there is no attachment of males to particular sites. The similarity with lekking species extends to the marked sexual dimorphism shown by many duck species. Because a male's success is dependent on his being attractive to females, sexual selection has favoured the evolution of bright plumage and elaborate displays in males. Male common eider ducks *Somateria mollissima* perform no fewer than 13 different displays during courtship.

Where the mallard's mating system differs from that of lekking species is in the fact that the mallard is monogamous.

By the early spring pairs have been formed and these disperse from the mating site, migrating to areas of suitable nesting habitat. The male stays with the female and defends her until she has finished laying her eggs. He then leaves her to incubate her brood alone and joins up with other males to form small groups. At this stage, males will attack and attempt to mate with females who have not yet laid all their eggs, though they will have their victim's male to contend with. These mating attempts by males are strongly resisted by females and are referred to as rape. In some ducks females have been known to be severely injured and even killed by raping drakes.

Mallard reproductive behaviour thus incorporates different elements that we would not expect to find within a single mating system. The social display groups, with their strong element of aggressive male competition, the very marked sexual dimorphism, and the fact that females alone incubate the eggs, are all features normally associated with polygyny. In that a female only mates with one male, the mallard is clearly monogamous, though the only period when male and female act as a pair is when the nest is being built and the eggs laid. That competition between males occurs in this monogamous species is partly due to the fact that there are fewer females than males, probably because females suffer higher mortality during the nesting and incubation phases of the breeding cycle. The picture is further complicated by the fact that rape occurs. Rape does lead to fertilization so that a few of the eggs in a clutch laid by a female who has been raped will be fathered by a male different from the one she originally paired with.

While the mating system of the mallard and of other ducks is difficult to categorize because it contains elements of different mating systems, other species are difficult to classify because they may show different mating systems

Social display. A mallard drake performing a wing-flapping display to a female. Intense competition between males to attract females has led to the evolution of vivid male plumage and a number of conspicuous displays.

in different situations. We have already seen that in the long-billed marsh wren some males are polygamous while others are monogamous. The incidence of polygamy is higher in this species when food resources are very clumped than when they are dispersed. In spotted sandpipers *Tringa macularia* in North America, polyandry is prevalent where population is density is high and rare where it is low. What is important in understanding the biological basis of different mating systems is not whether or not we can fit each species into a convenient category but the extent to which we understand the factors that control mating systems. In this respect, far from confusing the picture, species with variable mating systems have given us considerable insights into the role of ecological factors in the evolution of mating systems.

4 Finding a mate

Before any courtship and mating activity can take place, animals have to find potential mates. The location of mates has two important biological aspects. A partner must be found who is both of the opposite sex and of the same species. Many features of the sexual behaviour of animals are adaptations for finding a potential mate, often over a considerable distance and in the face of many difficulties, and for correctly identifying its sex and species.

Mates which never meet

In some species mating effectively occurs without the partners ever meeting. Many sessile aquatic animals like *Hydra* and sea anemones simply shed their eggs and sperms into the water and whether or not they meet depends on the vagaries of water currents. A great many sessile animals are hermaphrodite and often they avoid fertilizing their own eggs only by producing eggs and sperms at different times. One sessile marine animal, the barnacle, has evolved a less fortuitous method of mating. Barnacles are hermaphroditic and each individual possesses two enormously elongated penises. A barnacle whose body is only half a centimetre in width has penises each of which can extend over 20 centimetres from the body. The penises are protruded and each is moved about over the surrounding rock until it contacts another barnacle. The penis is inserted into the shell of the partner and sperm is emitted. The partner may extend one of its own penises and reciprocal fertilization then occurs.

There are some very mobile animals that mate without meeting one another. Tiny male springtails move about among leaf-litter producing sperm in little packets called spermatophores which are deposited on the top of stalks. Female springtails walk among these and, if one brushes against a spermatophore it breaks open, releasing the sperm which may then find their way to her genital opening.

Mating in groups

Finding a mate is not a problem for animals that habitually live in social groups. Nor is it for those that gather to breed colonially at traditional sites. Many sea birds, like shearwaters and albatrosses, lead essentially solitary lives for most of the year but gather in large numbers to breed on remote islands. The entire world population of Abbott's booby *Sula abbotti*, a member of the gannet family, gathers to breed at only one locality, on Christmas Island in the Indian Ocean. For species like these the major problem is not locating a mate but finding the way back to the breeding site. Such birds have remarkable powers of navigation, and return to the colony where they hatched

Above An Abbott's booby and its chick. Left Unable to move, a barnacle extends a greatly elongated penis into the shell of a neighbour.

after perhaps two or three years of wandering vast distances over the ocean.

In species like ruffs that mate at leks there is often a strong element of tradition in the siting of a lek. Animals may find their way back each breeding season on the basis of their experience in previous years. In addition, the frenetic activity that typically occurs at leks provides a conspicuous target for animals unfamiliar with the area who are looking for a mate. The prairie chicken *Tympanuchus cupido* of North America was once very numerous but is now nearly extinct. In its heyday it

formed very large leks at which males
postured and produced loud booming
calls by alternately inflating and de-
flating large, bright-orange air sacs,
one on each side of the head. The combined sound of hundreds of booming
males was such that 'the very earth
echoed with a continuous roar' that
could be heard several miles away.

Leks are only one situation in which
the combined behaviour of many
animals makes it easier for individuals
to find mates. In some species of
firefly, swarms of light-emitting males
are so large and bright that they have
been used as beacons, at night, by
human navigators. In many of those
species of frogs and toads that form

explosive breeding aggregations, the
males emit loud calls. The resulting
chorus acts as an auditory beacon, for
other males and females, which in
some species can be heard over a mile
away. In this respect there is an interesting difference between the European common toad *Bufo bufo* and many
other frogs and toads. Toads typically
breed in deep ponds which are permanently full of water. Their eggs and
tadpoles are distasteful and are not,
therefore, attacked by fish and other
aquatic animals. They show very strong
mating-site fidelity, returning each
year to breed in the same pond. Toads
have been known to return to a site
where they had previously bred but

which had since been covered over by a new road. Male toads have a mating call but it is uttered very rarely and can only be heard over a range of a few yards. By contrast, two species in Europe, the common frog *Rana temporaria* and the marsh frog *Rana ridibunda* tend to breed in shallow, temporary ponds where their unprotected eggs and tadpoles will not be subject to predation by fish and other animals that require permanent water. The location of such ponds may change from year to year and frogs may appear in the spring at any newly-formed patch of water. They have loud mating calls and often form spectacular choruses. Since toads use dependable, traditional breeding sites they do not need a mating call to attract other toads in the vicinity. Each toad 'knows' where it is going. Conversely, it is advantageous for frogs to set up a chorus to attract other animals who are looking for a breeding site.

Another way to find a mate is to take up a position near food or some other resource which is essential to members of the opposite sex and which is therefore likely to be frequently visited by them. The males of many pollen- and nectar-gathering insects hover around flowers and wait for females to come and feed. Female dungflies *Scatophaga stercoraria* fly to newly deposited cow-pats where they will mate and lay their eggs. The dung, which provides the environment for the developing eggs and larvae, is the focus of mating activity. The females find new cow-pats by detecting their smell and thus tend to arrive at a new pat from a downwind direction. Male dungflies gather around new pats and concentrate near the downwind side where they are more likely to encounter a virgin female before another male does. In some species males compete for the possession of the resource to which females are attracted and a territorial system may result. This is the basis of resource-based polygyny which was discussed in the previous chapter.

Location by sight, sound and smell

Many species of animals show patterns of behaviour which are clearly adapted for attracting potential mates over a long distance. These advertisement displays may be visual, auditory or olfactory, depending on the senses that are most highly developed in a particular species. Visual advertisement displays in a number of nocturnal insects take the form of light which is emitted by special luminescent organs in which a light-producing chemical reaction takes place. In the glow-worm *Lampyris noctiluca*, a continuous light signal is produced by females positioned among foliage near the ground. In fireflies, males produce a series of flashes as they fly around. A female waits in vegetation and responds to the flash pattern of a male of her own species by producing a distinctive series of flashes of her own. The male sees this and flies down to her. Females of the same species may congregate in groups and thereby amplify their signalling. While birds make great use of sounds to attract mates, they also perform a variety of visual displays. One of the most dramatic is that of the great bustard *Otis tarda*, a bird that lives only on very large expanses of open plain in Europe and Asia. It forms mating groups in the breeding season. Males do not produce any sound during the display in which they appear to turn themselves inside out. Normally brown, grey and black in colour, they cock their tails forward and turn their wings back and over to display a billowing mass of pure white feathers so that they look like a huge snowball on legs. Seen from a distance a group of displaying males shows up as a series of white flashes. This use of a visual rather than an auditory display may be an adaptation to the fact that sound does not travel far over an open, windswept plain.

Auditory mate location signals are a feature of a number of insects which can produce remarkably loud sounds in view of their small size. Crickets and

Overleaf Visual display. A male great bustard displaying to a female. The male is puffing out his chest and turning his wings and tail forward to expose a billowing mass of white feathers. This very conspicuous display is visible from great distances over open, grassy plains.

grasshoppers produce sound by stridulation, a process in which two limbs are rubbed together. Grasshoppers rub the top of their hindleg, which is equipped with a series of raised processes, against a vein in the forewing which vibrates as a result. Crickets rub their two forewings together. Even louder than the chirps and trills of crickets and grasshoppers are the incessant 'calls' of cicadas. These are produced by a pair of special organs on the abdomen, each consisting of a tightly stretched membrane. The membrane is pulled inwards by a set of muscles and then released,

producing a click sound. These clicks are produced at such a rate, 100 to 500 a second, that they are heard as a continuous whine. In all these sound-producing insects, it is the male that advertises himself and the female who moves towards the source of the sound. The same is true of many species of frogs and toads, in which males may produce a variety of different calls that fulfil different functions. One of these, which attracts females, used to be called the mating call but is now generally referred to as the advertisement call because, as well as attracting

Below When calling from the breeding pond, the male American great plains toad *Bufo cognatus* inflates a large vocal sac beneath his chin which amplifies the sounds produced by his vocal cords.

Left A ghost crab
uses his environment
to attract females.
The male has dug a
burrow and piled the
excavated sand into
a pyramid which the
female can see from
a far greater distance
than she can see him.

females, it may also function as an inter-male territorial signal, causing males to keep away from one another. Many male frogs possess vocal sacs which are inflated with air and which amplify the sound that is produced by vibrations of their vocal cords. Some species have a single sac beneath the chin, others have a pair of sacs, one on each side of the mouth. A similar adaptation is seen in the sage grouse *Centrocercus uropha-sianus*, a lekking bird in which the male has an inflatable throat pouch which amplifies a booming sound produced during lek display.

Some species make ingenious use of their physical environment to increase the effectiveness of their visual or auditory signals. Male bower birds, of the Australian region, do not merely rely on their bright plumage but build a special bower on the ground. This bears no relation to the nest which is built in a tree and may be hundreds of yards away. In front of his bower a male has a display area which he decorates with a carefully selected variety of coloured or highly reflectant objects. The kakapo *Strigops habroptilus* is an extremely rare and aberrant species of parrot that is found in New Zealand. It has virtually lost the power of flight, lives almost exclusively on the ground and is nocturnal in its habits. In the breeding season it forms leks and males advertise them-

selves by producing a booming sound rather like that made by blowing over the neck of a bottle. A male may boom a thousand times an hour for six or seven hours a night. They are especially vociferous on damp, windless nights when their sound travels particularly well. The sound is amplified by filling air sacs within their bodies, making themselves almost spherical in the process, and by calling from the edge of a hollow which they have excavated in the ground. It used to be thought that these 'bowls' were dust baths, but it is now known that they are positioned and shaped in such a way that they reflect the sound far and wide over the surrounding countryside. In the old-world mole-crickets *Gryllotalpa vulgaris*, the male stridulates inside a burrow in the ground. This has a characteristic Y-shape; the two openings to the air at one end make it act as an amplifier thereby increasing the range of the cricket's 'call'. Ghost crabs of the genus *Ocypode* are further examples of animals that use their burrows to advertise their presence to females. These crabs are widely distributed throughout the sandy beaches of the tropics, and the male builds the excavated sand from the burrow into a pyramid to which the female is attracted.

In olfactory communication, animals convey information to each other by chemical 'messengers', secre-

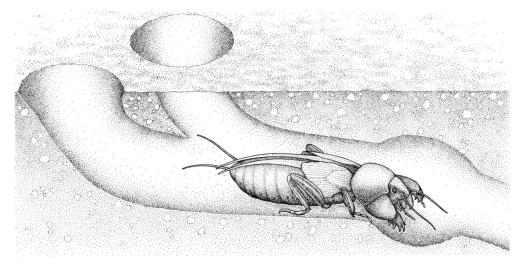

Right The distinctively shaped burrow of the male mole-cricket acts as an amplifier that increases the range of his mating call.

Above The huge antennae of the male Tussore silk moth are organs of smell which enable him to detect the distinctive odour of the female at a range of several miles.

Left A male spotted bowerbird *Chlamydera maculata*. To increase his attractiveness to females, he has spread a selection of bright objects, including a toothbrush, in front of his bower.

tions called pheromones that are produced in special glands. Some pheromones are effective only over a short distance, either because their molecules are too large to be carried far or because they quickly break down. Others are highly volatile and very stable and can be detected several miles from their source. While olfactory communication is very important in the sexual behaviour of many mammals, the most highly developed olfactory mate location systems are found among insects, particularly moths. The female silk moth *Bombyx mori* produces a volatile secretion called bombykol from scent glands on her abdomen, and disperses it into the air by fluttering her wings. Carried by

the wind, the female odour can be detected by males as far as seven miles away. Males have enormously enlarged antennae carrying many thousands of sensory cells which are so sensitive that they will each respond to one molecule of bombykol. In response to the bombykol, the male flies upwind until he is able to locate the female accurately by moving towards higher concentrations of her scent.

Fish and marine mammals have evolved systems for finding mates in an environment which is often only thinly populated by their species. In water close to the surface, enough light penetrates to allow colour to be used as a sexual attractant, but in deep water, where the light intensity is too low for this, some species of fish, such as the lantern fish, produce light themselves. When the location of a mate is difficult, it is important that any meeting between the sexes is fully exploited. In angler fish of the genus *Photocorynus* the comparatively minute male becomes permanently attached to the female. Although he lives parasitically on his mate, the greatly increased chances of the female's eggs being fertilized offsets the disadvantage to her. Some fish employ olfactory signals to attract a mate; the courtship of the blind goby *Bathygobius* is initiated entirely by odours produced by the female. Communication by sound between the sexes is common in some families of fish, and is widespread in many marine mammals. Many pelagic fish and mammals congregate in schools or shoals before moving to a suitable breeding site. The shoal presents an easier target for each individual to find, and increases the chances of each finding a mate in breeding condition.

In advertising themselves to prospective mates, animals that send out long-range signals are making themselves conspicuous not only to members of their own species but also to others who may be predators or parasites. In some species, the risks inherent in sexual advertisement are considerable. North American bullfrogs *Rana catesbeiana* gather to breed in ponds where males establish territories, calling frequently to attract females and to proclaim the ownership of their territories to other males. The breeding pond may also be inhabited by snapping turtles *Chelydra serpentina* who eat frogs and who locate them by listening to their calls. A species of frog *Eleutherodactylus fitzingeri* that lives in Central America is preyed on by the fringe-lipped bat *Trachops cirrhosus* which swoops down and takes calling males in its mouth. The frog appears to have become adapted to this risk of predation; this species calls much less

frequently than other, poisonous species not exposed to the same threat. An edible male frog is faced with a serious dilemma; he must call often enough to attract females, but in doing so runs the risk of being eaten. Male fireflies of the genus *Photinus* run the risk of being deceived and eaten after they produce their sexual flash display. Females of another firefly genus, *Photuris*, mimic the female flash patterns of the *Photinus* species. A *Photinus* male, responding to what appears to be a female of his own species, flies down and is eaten.

A more complex relationship exists between a North American cricket, *Gryllus intiger* and a parasitic fly, *Euphasiopteryx ochracea*. Male crickets gather in groups and form a chorus. Within the group there are some males who do and some who do not call; females are attracted to the group by the songs of the callers. Non-calling males attempt to intercept and mate with females that are moving towards the callers. In terms of short-term mating success, callers fare better than non-callers since females are directly attracted to them. Furthermore males who call loudly attract more females than those with soft calls. However, the calls also attract the parasitic fly which comes near to a group of calling males and then deposits its larvae. The larvae are sticky and adhere to any cricket they come into contact with. They then eat their way into the cricket, eventually killing it. Non-calling male crickets are much less likely to be infested with fly larvae than are calling males. The cricket provides an example of behavioural polymorphism, which is the existence of more than one genetic type in the population. Males are either callers or non-callers throughout their lives. The polymorphism is maintained by the balance of advantages and disadvantages experienced by the two types of male. Callers are at an advantage over non-callers in that they mate more often but are at a disadvantage be-

Opposite The cost of sexual advertisement. A male frog, *Eleutherodactylus fitzingeri*, calls and attracts the attention of a fringe-lipped bat which swoops down to eat him.

cause they are more likely to be killed by fly larvae.

In the great majority of these and other examples of long-range mate attraction behaviour, it is the male who advertises and the female who responds. The silk moth, like other species of moth, provides an exception to this rule. Why should it be that it is the male who usually advertises himself? This question can be answered in terms of the concept of reproductive effort that was discussed in Chapters 2 and 3. Females produce fewer gametes than males, devote a larger proportion of their reproductive effort to parental care and are essentially a limited resource for which males must compete. Among males, who devote much of their reproductive effort to mating, there is powerful selection favouring those which are most successful in competing for or attracting females. This selection has resulted in the often sophisticated and elaborate displays which males employ to attract females, often in the face of considerable risks from enemies. It may be significant that in the case of moths like the silk moth the signalling female does not appear to be at any risk from predators. This is because she produces a highly specific signal in the form of a single chemical compound to which only males of her own species are receptive. In effect she is advertising herself on a completely safe channel to which no other species are tuned. Animals using visual or auditory signals are broadcasting on channels to which many species, including enemies, are tuned.

We have seen examples where one species, by mimicking a sexual signal of another species, can deceive members of the other species for predatory purposes. This is called aggressive mimicry. A non-aggressive form of mimicry that also involves deception is shown by some plants whose flowers are so similar in appearance to females of particular insect species that males of these species attempt to mate with them. In so doing the males pick up the flower's pollen and thus unwittingly aid the plant's reproduction. Some of these flowers have been found with insect sperm inside them.

Choosing the right sex

For animals that are in frequent contact with each other, a problem that may arise in relation to mating is that of differentiating between males and females so as to ensure that sexual behaviour is directed only towards members of the opposite sex. In many species there is marked sexual dimorphism and the sexes are readily distinguished. In many bird species, the females are drab while the males are colourful and possess extra plumes, crests or enlarged tails. Often these differences are accentuated in the breeding season; at other times of the year the two sexes are more or less alike in appearance. When not displaying, it is better to remain camouflaged. In some species, as we saw in the previous chapter, differences between the sexes have often been exaggerated far beyond what seems necessary for sex recognition, by the influence of sexual selection. However, in many species, males and females are very similar to human eyes and are differentiated by very subtle differences in appearance, odour or behaviour.

Male and female North American checkered white butterflies *Pieris protodice* look identical to our eyes. However, they look very different to each other because their wings contain different amounts of a pigment which absorbs ultraviolet light to which these butterflies are receptive. Males have more of this pigment than females and so look darker to other butterflies. If the amount of pigment in a male's wings is reduced by treatment with a chemical extractor, that male becomes attractive to other males who respond to it as if it were a female.

Surprisingly, when females, who have small amounts of pigment in their wings, are treated in the same way, they

A male common toad *Bufo bufo* will clasp anything.

become less, rather than more attractive to males. It appears that males are not attracted to butterflies with very large or very small amounts of pigment but are attracted to those with intermediate amounts. The adaptive value of this preference is related to the fact that the pigment fades with age so that young females are slightly darker than older ones. Young females are more likely to be virgins and therefore to be carrying more unfertilized eggs than older females, who are likely to have mated several times. Thus a male checkered white butterfly can tell

from the intensity of ultraviolet light reflected from another butterfly's wings not only its sex, but also its age.

In the mêlée of a frog or toad breeding pond, it is quite clear that males have very poor powers of sex discrimination. They will clasp anything that moves and male European common toads *Bufo bufo* have been observed in amplexus with a variety of inappropriate objects such as other males, frogs, goldfish, the handle of a net and waterweeds. A male toad will clasp anything of a suitable size, especially if it moves. The most common moving objects in a toad

A male red-breasted merganser points skywards with his beak and paddles with his foot. This display shows off his distinctive plumage to females.

breeding pond are other male toads who frequently clasp each other. When a male is clasped he gives a characteristic call called the release call to which the clasper responds immediately by letting go. Male toads are thus able to recognize females because they do not give the release call when clasped.

Choosing the right species

In locating a mate it is biologically vital that animals find partners who are not only of the opposite sex but who are also members of the same species. Sexual activity between members of different species may have no more harmful effect than wasting time that could be devoted to courting and mating with members of the same species. At worst, it can lead to the total loss of an individual's reproductive potential. The reason why mating between species is biologically harmful is that, because species differ in their genetic make-up, fusion of gametes from two species very rarely produces combinations of genes which are viable, meaning that they are capable of surviving to reproduc-

tion. The nature of this effect, called hybrid breakdown, varies from one combination of parental species to another. At one extreme, the sperms of one species may simply fail to fuse with the ova of another. At the other extreme there are hybrid matings that produce fully viable, fertile progeny, but these are unable to reproduce successfully because their genetic make-up is incompatible with that of either the parental species. Between these extremes there is a gradation of effects, including death at some stage in embryonic development, hybrids that develop fully but which are less viable than pure-bred progeny and which therefore die early in life, and hybrids that are fully viable but which are totally sterile. The mule, for example, is a fully viable but sterile hybrid between a horse and a donkey.

An interesting kind of disadvantage to which hybrid offspring may be subjected is shown by hybrids between the American tree-frogs *Hyla chrysoscelis* and *Hyla femoralis*. These are viable and reach sexual maturity, but the male hybrids produce a mating call that is intermediate in many of its characteristics between the male calls of the two parent species. It has been shown that pure-bred females, given a choice between the call of a male of their own species and that of a hybrid male, always prefer that of their own species. The hybrid males thus have very low reproductive success because of their inability to attract females.

Hybridization is especially important in the relationship between closely related species. Usually, such species have recently evolved from two populations of a common ancestral species which became separated by some geographical barrier like ice sheets or sea. If for some reason their ranges come together again, such species are likely to be very similar in their ecological requirements. Individuals belonging to both species are likely, therefore, to encounter each other in the breeding season. They are also likely to be similar in appearance and morphology so that misidentification of species will tend to occur and hybrid mating may be anatomically feasible. In view of the severe disadvantages that may be suffered by hybrids we would expect natural selection to have favoured the evolution of adaptations that accentuate differences between closely related species, especially in characters related to mate recognition. This is indeed what we find.

Most species of ducks are highly sexually dimorphic in appearance. Females are generally brown and speckled whereas the males are brightly coloured. Comparing the different species of ducks, the females are very similar in

appearance and are often difficult to distinguish in the wild, whereas the males have very different plumage patterns and colours and are very distinctive. The brilliance of male plumage in ducks can largely be attributed to the fact that it is the males who take the more active role in courtship, eliciting sexual responses from females by a variety of elaborate displays. But it is the females that choose the males and so sexual selection has presumably favoured males with the most conspicuous plumage patterns. The differences between species in male plumage and displays can be interpreted as the result of selection favouring those males whose appearance is most unlike those of

Animals do make mistakes. This male Rio Grande leopard frog *Rana berlandieri* is clasping a male bull frog *Rana catesbeiana*. He has thus misidentified both the species and sex of his partner.

closely related species. There has thus been divergent evolution in male plumage characteristics. By contrast, the similarity of female plumage across species is due to convergent evolution favouring plumage that makes females, who incubate the eggs, well camouflaged as an adaptation against such predators as foxes. Since it is the females who exercise the choice, there is no selection for difference in their plumage.

There are many species of firefly, some of which live in the same localities and perform their courtship flights on the same nights. The males of the different species produce distinctive patterns of flashes, each consisting of a stereotyped series of short or long flashes. Females respond only to the male flash pattern that is characteristic of their own species.

It is commonly observed that males are very much less discriminating with respect to species in their sexual behaviour than are females. We have already seen that male toads will clasp a very catholic variety of objects, including frogs of both sexes. Males of some of their amphibian relatives, the newts *Triturus*, display to females of closely related species as vigorously as they do to females of their own species. Female newts, however, will only show positive responses to males of their own species. Why should males be less dis-

criminating than females? One possible explanation is that if females are discriminating in mating there is no selective advantage in males being discriminating, because there will be very few hybrids against which natural selection could act to favour male discrimination. Alternatively, one can argue that females have more to lose than males by participating in hybrid matings and are therefore subject to more powerful selection. This is because females frequently put all their reproductive effort into one mating and can thus waste it all by making one mistake. Males may have many opportunities to mate and, provided that they make some correct matings, they can afford to make a few mistakes.

Some of the clearest and most elegant examples of the relationship between sexual behaviour and the maintenance of species integrity are provided by frogs. The typical pattern of sexual behaviour in frogs is that males produce calls that are species-specific and that females approach them, responding selectively to the call appropriate to their species. Thus hybridization between closely related species is prevented both by the specificity of the male's call and by the selectivity of the female's response. Some species of frogs can only be differentiated on the basis of their calls. Two North American

Recognizing the correct species by means of a code. Males of different firefly species produce different patterns of flashes as they fly around. The females, from their position on vegetation, respond only to the flash pattern appropriate to their own species. From the top, *Photinus brimleyi, P. collustrans, P. ignitus* and *P. granulatus.*

species were for a long time considered to be a single species, the gray tree-frog *Hyla versicolor*. However, some 20 years ago it was shown that there were two types of male, one whose call consisted of a rapid trill, the other a slower trill. Females could also be divided into two types on the basis of their behaviour in a test in which they were presented with a choice between recordings of the fast and of the slow call. Some females always approached the fast call, others the slow call. Experiments conducted to test the genetic compatibility of the two call-types found that if males and females belonging to different call-types were mated, they produced fewer offspring than did animals mated with partners belonging to the same type. Furthermore, the offspring of matings made across the call-types were less likely to survive than the offspring of matings made within call-types. Clearly, the two call-types are not genetically compatible and the fast-calling type is now given the status of a separate species, *Hyla chrysoscelis*. The slow-calling type retains the original species name.

This example shows that, during evolution, closely related species can become more different in their sexual behaviour than in other aspects of their biology. Clearly there must have been very powerful selective forces at work to have produced such a marked

difference. These selective forces arise when hybridization occurs. As we have seen, hybrids are typically less viable or fertile than pure-bred offspring, so that individuals who participate in hybrid matings are very unlikely to pass their genes on to subsequent generations, whereas those mating within their species will. This substantial difference in reproductive success between those mating between and within species provides the powerful selective force that favours any characteristic that increases an individual's chance of mating only with a member of its own species. This selection will favour those males whose calls are most obviously characteristic of their species and will favour those females who are most selective in their response to males. These effects are shown very clearly in southeast Australia, where the ranges of two species of tree-frog, *Hyla ewingi* and *Hyla verreauxi*, overlap. Males of both species produce a trill-like call consisting of a rapidly repeated series of sound pulses. Males of the two species taken from outside the overlap zone have calls with very similar pulse rates, and females of the two species are not very good at discriminating between them. However, within the overlap zone, males have very distinctive calls, because in this area, *Hyla ewingi* have slightly slower-pulsed calls than over

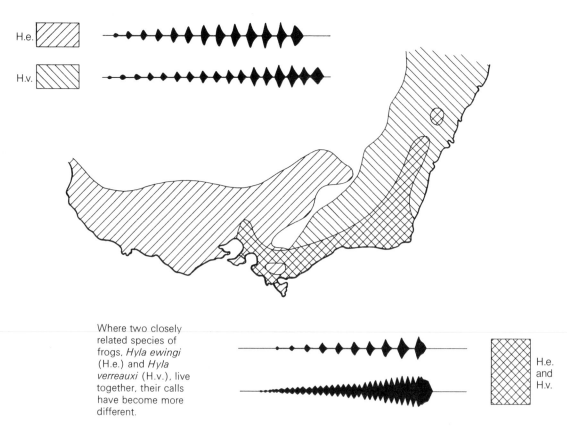

Where two closely related species of frogs, *Hyla ewingi* (H.e.) and *Hyla verreauxi* (H.v.), live together, their calls have become more different.

the rest of their range, whereas *Hyla verreauxi* has a call which is much faster in the overlap zone than elsewhere. In other words, the calls of the two species have diverged in the area where they occur together and where hybrid matings could occur. As one might expect, it is not only the male calls that have become modified in the overlap zone; so have the preferences of females of the two species. Females from the overlap zone are more responsive to extreme call rates, slow in *Hyla ewingi*, fast in *Hyla verreauxi*, than they are to the call rates that they would hear if they lived outside the overlap zone.

Female frogs that must discriminate between the male calls of their own and of other species may be faced by a rather special kind of problem. This is related to the fact that frogs, like all amphibians, are not able to maintain a stable body temperature in the way that a mammal, a bird and, to a lesser extent, a reptile is able to. Thus, as air temperature changes, so does a frog's body temperature. Some features of frog calls are altered by changes in body temperature. While the sound frequency, or pitch, of a male's call is not affected by his body temperature, the rate at which he produces pulses of sound increases as his temperature rises. A warm frog produces a faster trill than a cold one. The implications of this effect in the American tree-frogs *Hyla versicolor* and *Hyla chrysoscelis* are as follows. At a given temperature *chrysoscelis* always has a higher pulse rate than *versicolor*. However, a male *chrysoscelis* with a body temperature of 14 °C calls with the same pulse rate as a male *versicolor* whose body is at 26 °C, *ie* at about 25 pulses per second. There is thus every probability that a female listening to two such males would be unable to discriminate between them. She can resolve this problem if she takes into account her own temperature which is likely to be very nearly the same as that of any two males calling in her vicinity, and this is exactly what she does. Given a choice between two calls that differ in pulse rate, a female approaches the one whose rate is most like that of a male of her own species calling as if he were at a temperature the same as hers. Thus, on warm nights, when male calls become faster, the female's preference also shifts towards faster calls. Because a female's ability to discriminate different pulse rates is coupled to her own temperature, it is not impaired by changes in the environment that alter the nature of the male call.

We have seen in this chapter that the task of finding a mate is often not a simple one, but one which is beset by difficulties and hazards. For an animal the problem of finding a mate, sometimes over a great distance, is compounded by the presence of enemies who may be tuned into its communication system or by the risk of being attracted to and mating with members of other species. However, even when an animal has overcome these problems and located a mate of the same species, it may still have much to do before it can proceed with the act of mating. As we saw in Chapter 1, a fundamental aspect of female sexual strategy is the element of mate choice. While male reproductive success is often simply a matter of obtaining several mates, females tend to increase their reproductive success by mating selectively. For a female, it may not simply be a matter of finding a male of her own species, but of finding the best male. In Chapter 2 we saw that in some species the reproductive roles of the two sexes may be such that it is adaptive for males also to exercise mate choice. In the next two chapters we shall examine the element of choice in animal sexual behaviour and the competition that choice exercised by one sex can generate between members of the other sex.

5 Choosing a mate

Of the many members of the opposite sex that they encounter during a breeding season, individual animals usually mate with only one or a few of them. The question we examine in this chapter is whether this restriction of mating, to only a few of the many potential mates available, is based on some kind of choice. There are good reasons why we might expect animals to choose their mates. Individual members of a population differ from each other in countless ways. Some will be stronger, more fertile, more experienced or more brightly coloured than others of the same sex. An animal that somehow mates selectively with a partner who is superior to other potential partners in some characteristic relevant to survival or to reproduction, may thereby increase its own reproductive success. What the relevant characteristic might be will vary from species to species. For example, in a species in which males defend territories containing food or some other vital resource, a female may increase her reproductive success by mating with a male holding a high quality territory. It is also possible that, if the male's superiority has a genetic basis, a female, by combining his genes with hers, may pass his superior qualities on to her offspring, thus increasing their chances of surviving and breeding in their turn. Expressing it in more general terms, we would expect natural selection to favour any mechanism that enables animals to choose as mates the fittest members of the opposite sex.

When we use the word 'choice' in relation to the behaviour of animals we do not necessarily imply that they make rational or conscious choices as man might do when presented with several options. Mate choice is a term that includes any mechanism that increases the probability that an individual will mate with certain available partners rather than others. As we shall see, mate choice in animals takes forms that can be very subtle and indirect and which are not at all like the decisions which we refer to when we use the word choice in everyday speech.

The evolution of elaborate male plumage

Mate choice in animals has major evolutionary implications and is fundamental to the theory of intersexual selection which was discussed in Chapter 1. If all, or a large number, of females in a population choose as mates males with a particular characteristic, those males will enjoy greater reproductive success than males lacking that characteristic. The variation in male success that results from female preferences for certain males imposes a powerful selection favouring those males and leads to the rapid evolution of the male characters on which the female preference is based. For example, if females prefer brightly coloured males, selection will favour the evolution of bright male coloration and may lead to very exaggerated sexual dimorphism.

Since Darwin first proposed his theory of sexual selection there has been considerable debate about how intersexual selection might actually work. This has centered on the question of why females should choose males with such characters as bright coloration or long tail feathers. On the face of it, such a preference does not yield any obvious tangible benefit to a female. The most widely held view is that, in the initial stages of its evolution, a character preferred by females must have been associated with some benefit that accrued to females or to their offspring. For example, it would be adaptive for a female bird to choose a male with a long tail if having a long tail made the male a more effective or skillful flier, since she might pass this advantage on to her sons. Once the female preference for long-tailed males has become established in a population because of the benefits it confers, a new factor becomes important. This is that long-tailed males now have a double advantage. Not only are they better fliers, but they are also more attractive than short-tailed males. Over a number of generations this second advantage becomes more and more important. Females who choose attractive males benefit because they tend to have attractive sons who will therefore have an advantage in attracting females in their turn. Thus, once a female preference is established, natural selection will favour male attractiveness for its own sake. In each generation the males with the longest tails will attract the most females and so male tail length will tend to increase over several generations. This process will continue even if the increasingly long tail, originally an advantage in flight, becomes a liability

and decreases the chances of a male's surviving, provided that the advantage a long tail gives a male in terms of mating success is greater than the disadvantage he suffers in terms of survival. Eventually a stable point is reached where further elongation of the tail no longer confers a mating advantage sufficiently large to offset the reduced chances of survival. It is this kind of process that is believed to have led to the evolution of the incredibly elaborate and colourful plumage of peacocks and male birds of paradise. There can be little doubt that these are an enormous encumbrance to males in their day-to-day activities and that they must make them very conspicuous to predators. They can only have evolved if these disadvantages are counter-balanced by their effectiveness in attracting females.

We saw in Chapter 3 that in certain species, such as the huia and some birds of prey, sexual dimorphism may be unrelated to sexual behaviour but is an adaptation to ecological factors. A similar suggestion has recently been made in relation to very elaborate male plumage in birds. The 'unprofitable prey' theory suggests that conspicuous male plumage has evolved as a means of communicating to potential predators. As a general rule male birds that are brightly coloured or that have elaborate plumage tend to live in open habitats and to be very mobile and consequently very difficult for predators to catch. Predators will quickly learn that conspicuous birds are hard to catch and will seek alternative prey. Conversely, females, who cannot be mobile because they have to stay on or near their nest and incubate their eggs, are dull and well camouflaged. This very interesting theory is not entirely satisfactory. One can equally well argue that brightly coloured males have had to evolve means of being elusive to reduce the risk that arises from their being conspicuous to predators. It is essentially impossible to decide whether this theory or the theory of intersexual

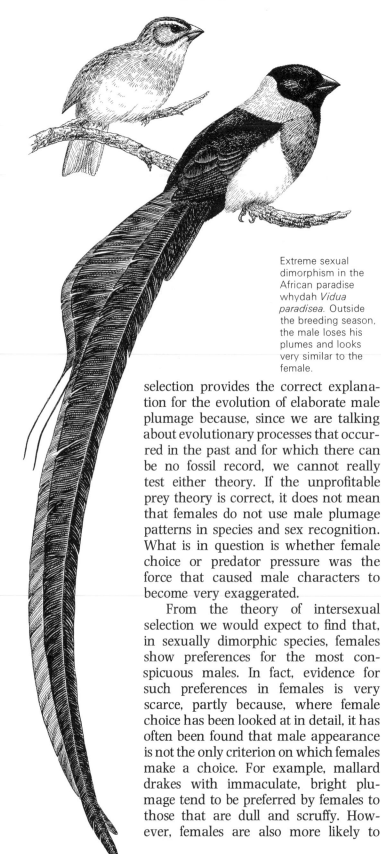

Extreme sexual dimorphism in the African paradise whydah *Vidua paradisea*. Outside the breeding season, the male loses his plumes and looks very similar to the female.

selection provides the correct explanation for the evolution of elaborate male plumage because, since we are talking about evolutionary processes that occurred in the past and for which there can be no fossil record, we cannot really test either theory. If the unprofitable prey theory is correct, it does not mean that females do not use male plumage patterns in species and sex recognition. What is in question is whether female choice or predator pressure was the force that caused male characters to become very exaggerated.

From the theory of intersexual selection we would expect to find that, in sexually dimorphic species, females show preferences for the most conspicuous males. In fact, evidence for such preferences in females is very scarce, partly because, where female choice has been looked at in detail, it has often been found that male appearance is not the only criterion on which females make a choice. For example, mallard drakes with immaculate, bright plumage tend to be preferred by females to those that are dull and scruffy. However, females are also more likely to

mate with the drakes highest in the male dominance hierarchy, who are not necessarily the most handsome males. Indeed, some of the most beautiful male mallards are birds who avoid fighting and who are therefore of lowest rank. Such males are not chosen by females. Among drakes who do fight, bright plumage may be an indicator of strength because only a male who wins all his fights easily will come away from them with his plumage undamaged. Thus the adaptive value of the preference of female mallard for the brighter drakes may be that it ensures that their mates will be strong, not simply attractive.

Female choice of other male qualities

Whether or not male adornments are important in themselves, or only as indicators of other male qualities, physical attractiveness is only one of many criteria by which females might choose males. Which criterion is relevant in a given species depends very largely on the role that the male plays in reproduction. In the North American bullfrog *Rana catesbeiana*, males defend territories in which the eggs and larvae develop. Older, larger males attract more females than smaller ones, but this is not because they are more attractive in themselves but because they have successfully competed for those parts of the pond that females prefer as egg-laying sites. These are places where, because the water is fairly deep, temperatures do not get high enough to cause abnormalities in the developing eggs. Furthermore, the sites preferred by females are less infested than other parts of the pond by a predatory leech that eats bullfrog eggs. Thus, in a species like the bullfrog whose mating system is a polygynous one based on ecological resources, female mating preference is based on the quality of resource that a male controls, not on his appearance or behaviour. In Chapter 3 we saw other examples of female choice based on the resources controlled by males. Female long-billed marsh wrens choose territories with a rich food supply and female dickcissels those with good cover for nests.

The male three-spined stickleback *Gasterosteus aculeatus* defends a territory which provides an area of relative safety in which eggs and young can develop and grow. The larger a male's territory, the less likely is it that eggs or young will be eaten by a predator, which is very often another male stickleback. The courtship of sticklebacks has been studied in great detail. Whereas most studies have focussed on the male's behaviour and have tended to assume that the female plays a rather passive role in courtship, a recent study has suggested that this is far from true. Females swim around in large, rather ill-defined home ranges which may encompass several male territories. Females are aggressive towards one another and a group living in a particular area establishes a dominance hierarchy. Dominant females gain more opportunities to enter male territories and participate in courtship. Subordinate females tend to be interrupted and driven away by dominant females if they start to court a male. The courtship of sticklebacks is a protracted affair, consisting of a series of interactions between male and female. This sequence may be terminated at any point by either sex, but the female is more likely than the male to break off from courtship, suggesting that she is selective about the males to which she will respond positively. The point in the courtship sequence when a female is most likely to terminate courtship is at the stage when the male leads her to his nest and adopts a head-down posture, apparently showing his nest to her. Females who swim away at this point have presumably rejected the nest as unsuitable for their eggs.

In species in which males gather harems and control access to females directly, a female may appear to have little opportunity to exercise mate choice. However, even in the fiercely aggressive mating system of the elephant seal, females can influence the outcome of competition between males. A cow will often call loudly when a bull attempts to mount her and she is more likely to do so if he is of low rank in the male hierarchy. Hearing her protests, a nearby dominant bull will come over and chase the subordinate away. Her behaviour thus increases the probability that she will be mated by a dominant male. By doing so she may increase the chances that her sons will inherit any genes that contribute to male strength.

In lekking species, females often seem to select males not only on the basis of their displays or appearance, but show an apparently arbitrary preference for those males that hold certain traditional positions, which may or may not be at the centre of the lek. Since these are the object of the fiercest competition between males, the advantage that females gain from such a preference may be that they mate with the strongest males. In some species the best positions are held by the older males, possibly because they have the experience to know which positions have been preferred by females in previous years. Mating with an older male can be advantageous to a female because his ability to survive has been proven. A female may be able to pass

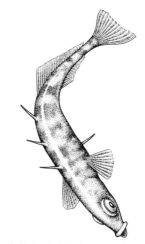

A male stickleback in the aggressive posture with spines raised.

genes that have contributed to the male's survival on to her offspring. In the black grouse *Lyrurus tetrix*, females select males not only on the basis of their position but also on their courtship behaviour. Older males attract more females than younger males, apparently because their greater experience of courtship makes them more skilful at adjusting their displays so as to stimulate females sexually.

In polygynous species the distribution of matings among males is primarily determined by inter-male competition. Females may, however, behave in ways that increase male competition thus making it more likely that they will mate with a winning male. In a polygynous species a female usually gets little or no assistance from the male in parental care. In some species she gains access to essential resources held by males, but in others, notably in lek species, all she gains from mating are the male genes that constitute half the genetic make-up of her offspring. By encouraging males to compete for her, a female may ensure that the male she mates with is the one with the best genes available. This argument assumes that a male's competitive ability is at least in part genetically determined. We have already seen how female elephant seals can increase the likelihood of their mating with a dominant bull by provoking fights between bulls. Females can also increase competition between males by gathering in groups and this may have been a factor in the evolution of social mating systems such as that of the mallard, and lekking of species.

In many species competition between males does not express itself in fighting but in vigorous display to females. As in aggressive species, females can encourage males to put greater effort into their mating activity by gathering in groups, since they force males to display to them in the presence of rivals. In guppies *Lebistes reticulata*, single males display at a slower rate in the presence of females than they do when there are also several males present. A feature of many polygynous species is the enormous length of time that males have to display before females will respond positively to them. Females appear to demand far more courtship from males than we would expect if the only function of sexual displays was to ensure the correct identification of a mate's sex and species. It appears that in many species females have evolved very high thresholds to male display and that males have consequently had to adapt by evolving more elaborate and sustained courtships. Males of the Pacific tree-frog *Hyla regilla*, form choruses in which several males call together. Calling occurs in bouts and individual males sustain calling for different amounts of time. Within any one chorus it is often possible to identify a chorus leader who is usually the first to start and the last to stop calling and who also calls louder and at a faster rate during a calling bout. Experiments conducted using recorded calls showed that females are more likely to approach a chorus leader than other males. What females gain by mating preferentially with males with the most vigorous and prolonged courtship displays is not certain. It is possible that the male who puts most effort into his courtship is the fittest male. This would be true if the ability to court depends on his ability to perform other vital activities such as finding food and avoiding predators. At present there is no convincing demonstration of any link between male ability to court and general fitness and so we must remain cautious about suggesting what benefits females may derive from being selective.

Even if females could somehow mate selectively with those males with the highest genetic fitness, there are serious theoretical difficulties in arguing that such selectivity could have

Opposite A brightly coloured male guppy displaying to a female gravid with eggs. The females are rather unresponsive and the males spend a lot of time displaying fruitlessly.

provided the basis of the sexual selection of male characters. If selection is very intense, that is if all the females in a population mate with only a few of the males, over a very few generations males will become very uniform in their characteristics due to a reduction in their genetic variability. This effect is a common experience in animal breeding programmes where attempts to breed a leaner strain of cow or a faster race horse produce marked improvements in the first few generations but very little subsequently. Very powerful selection effectively eliminates the genetic variation on which further selection could act.

What females may be doing by forcing males to display long and vigorously to them is testing them in some way. The European smooth newt *Triturus vulgaris* has a prolonged aquatic courtship sequence in which the male has to display to the female by means of a number of strenuous tail movements before she starts to respond to him. His ability to do this is constrained by the fact that, although he can take in a certain amount of oxygen through his skin from the surrounding water, he is primarily dependent on air obtained from the water surface. Thus his courtship is a kind of race between his being able to stimulate a female to the point at which she becomes responsive and his having to go up to the surface to gulp in a fresh supply of air, when he is likely to lose contact with the female. Thus, a female, by delaying her response may mate selectively with males who are able to hold their breath for a long time. Again, it is not at all clear what advantage she might gain by this.

In any species in which the male makes a substantial contribution to the parental care of the young, we might expect females to choose their mates on the basis of their qualities as parents. The male stickleback drives a female away as soon as he has fertilized the eggs she has laid in his nest and thereafter he alone tends the nest, fanning the eggs to keep them well oxygenated, defending them against predators and, when they have hatched, keeping the fry within the safety of the territory. Females show a preference for very aggressive males who defend large territories. These are more likely to be effective fathers because they are more likely to keep away other stickleback males, who readily eat the eggs and young of their neighbours. The terns *Sterna* are monogamous and both parents share in the protection and feeding of the young. During courtship, males catch fish and offer these to prospective mates. There is evidence that female common terns *Sterna hirundo* are more likely to pair with males who bring them many fish than those who only bring a few. It is possible that females thus choose males on their ability to catch fish, an ability that will be important to the survival of the young when they eventually hatch. There are significant correlations between the number of fish a male provides the female during courtship, the amount of food he brings to his chicks and the fledging success of his clutch. Thus his performance during courtship feeding is a reliable predictor of his potential as a parent.

Whether or not a female uses courtship feeding by a prospective mate as an indicator of his ability to provide food for her progeny, she certainly derives an immediate material benefit from the male's behaviour. She receives extra food at a time when she most needs it because she is developing a batch of eggs within her. It is a short-term benefit that seems to be the adaptive value of the mate choice shown by female hanging flies, *Bittacus apicalis* in North America. Hanging flies feed on other insects, and before they mate males fly around in search of prey. Having found a victim they kill it but do not eat it. Instead, they fly around in search of females, carrying the dead prey with them entangled in their legs which hang below the body.

Opposite Feeding the mate. A male little tern *Sterna albifrons* gives a fish to his mate who is incubating two eggs.

A male village weaverbird puts the finishing touches to one of his nests. His ability to attract females depends partly on his skill in nest building and partly on the age of his nests. Freshly-built nests are stronger.

Males present their prey to females who mate preferentially with those males that carry the largest prey. One advantage that females gain by such a preference is that, the larger the meal given to them by a male, the more likely is it that it will sustain her until she is once again sexually receptive and will be fed by another male. Between matings there is a period during which the female develops a new batch of eggs. The less she has to fly around during this period, the more of her energies she can devote to egg formation and the less risk she runs of flying into a spider's web.

While female common terns may assess the parental abilities of prospective mates on the basis of their capacity to bring them food during courtship, the female village weaverbird *Ploceus cucullatus* selects her mate according to his ability to perform a different aspect of parental care, nest building. These African birds are polygynous and each male builds several nests, at which he displays vigorously to females, hanging upside-down near the nest entrance and flapping his wings. The

males build nests by tearing fresh leaves into strips and weaving these into a large, hollow ball that hangs from a branch of a tree. In a captive population, studies of the responses of females to different males and their nests have shown that females are much more likely to respond to a male who is displaying at a newly built nest that is still green than to one whose nest is old and brown. Males who have failed to attract females to old, brown nests tear them down and start new nests. Simply painting a brown nest green does not restore its attractiveness to females. They appear to be assessing nests on the basis of their strength rather than their colour. As nests become older the leafy material of which they are made becomes dry and brittle. It is advantageous to a female to prefer a newly built nest because it is more likely to hold together for the full time that it will take for her eggs to develop and hatch. A female's preference for a certain male is in fact not based solely on the quality of his nest but also on the vigour of his display. Furthermore, it takes considerable practice for a male to become expert at building a properly constructed nest, and so the females' preference means that they are more likely to mate with older males. Females who have bred before also show a preference for males with whom they have previously mated.

Maintaining a pair-bond
In species that breed several times during their life span, breeding success may increase from one breeding season to the next as an individual becomes more experienced and more skilled in the various tasks involved in breeding. It will be advantageous for individuals of such species to mate with partners who have prior breeding experience. The domestic pigeon *Columba livia* is a species in which both parents look after the young and in which monogamous pairs are normally formed for life, though birds who lose their mates

find new breeding partners. Recent work has shown that pigeons of both sexes, when given a choice, will mate with an old, experienced partner in preference to a younger bird. However, this preference does not extend to birds who are more than seven years old who are discriminated against. Beyond the age of seven, pigeons start to show a decline in breeding efficiency. In the same way, the breeding success of the kittiwake *Rissa tridactyla* increases from year to year, not only as a result of each individual's accumulating breeding experience, but also because established pairs improve their performance as a cooperative unit. It is therefore advantageous for kittiwakes to remain with their partner of the previous season, provided that as a pair they bred successfully. Kittiwake pairs who fail to fledge young in a season tend to split up and form different pairings the following year.

In many monogamous birds, songs and postures that serve to attract mates continue to be performed for some time after a pair-bond has been formed and mating has occurred. It is generally assumed that this serves to maintain and to reinforce the pair bond throughout the breeding season. Male zebra finches *Taeniopygia guttata* found in the drier parts of the Australasian region, have a courtship song which shows considerable variation between individuals. In a study of captive zebra finches, the males and females in established pairs were kept apart for two or three days. While isolated, females were played tape-recordings of two songs, that of their mate and that of another male with whose song they were already familiar. They were much more likely to approach and sit near the speaker that was playing their mate's song than that playing the alternative song. This experiment shows not only that females can recognize individual males on the basis of their songs but also that they develop a preference for the song of their mates. Such a pre-

ference will tend to keep a pair together for the duration of a breeding season.

Avoiding incest

A factor that may be important in determining the reproductive success of an animal is whether or not it mates with a genetically related partner. Individuals that mate with a close relative, such as a sibling or a parent, generally produce less viable offspring than those who have unrelated mates. The harmful effects of incestuous mating are well known in man and form the basis of cultural taboos and legal restrictions that limit the incidence of mating between close relatives. It would clearly be adaptive for animals to be able to recognize their close relatives and to avoid mating with them. The existence of possible mechanisms for achieving such discrimination has been investigated in house mice *Mus musculus*. Olfactory communication is important in all forms of social interaction in mice, and this was observed in a study in which female mice were given pots containing sawdust collected from the cages of males that were their brothers and also from the cages of unrelated males. The females spent much more time sniffing the sawdust that smelt of the unrelated males, and this preference reduces the probability that female mice will engage in incestuous matings.

Male choice

So far in this chapter, we have looked almost exclusively at mechanisms of mate choice in females. As we saw in Chapter 1, the fact that females generally have a very limited reproductive potential in comparison to males means that we would expect selectiveness in sexual behaviour to be more common in females than in males. However, in species in which they make a substantial contribution to the care of the young, males also have limited opportunities to mate with several females. Accordingly we would expect to find that, in such species, males tend to be discriminating in their choice of mates.

Male sticklebacks show an ambivalent response to females who swim into their territories. A male's courtship display is frequently interrupted by sudden attacks in which he rushes towards the female and bites her. It is possible that this apparently counterproductive behaviour has evolved as a way of testing the sexual responsiveness of females. Females whose sexual motivation is high because their eggs are ready to be laid, may be less likely to swim away when they are attacked than those whose motivation is low. Though this theory is entirely speculative it may provide an evolutionary explanation for the rather puzzling occurrence of aggressive behaviour during courtship, not only in sticklebacks but also in a wide variety of other species.

Pigeons are normally monogamous for life and parental duties are shared, even to the extent that both male and female secrete the crop 'milk' on which the newly hatched young are fed. As we have seen, both sexes show a preference for older, more experienced partners. Age can be determined from plumage characteristics, but how pigeons assess the breeding experience of a potential partner is not certain, though it is thought that the courtship displays of experienced birds differ from those of naive birds. Male pigeons also show a very surprising discrimination against females who respond very eagerly to them during courtship. The sexual responsiveness of females can be raised by allowing them to be courted by males; such females tend to be rejected and even attacked when they are presented to a new male. Male pigeons display more vigorously to unresponsive females than they do to very responsive ones. This preference for females whose sexual motivation is apparently low is interpreted as an adaptation that decreases the probability that a male will mate with a female that has recently mated with another male. The advantage of such a preference is that it reduces the risk that a male will pair with a female, some or all of whose eggs have already been fertilized by another male. Putting it anthropomorphically, males who prefer unreceptive females are less likely to be cuckolded and thereby misled into caring for young of which they are not the father.

In this chapter we have seen that mate choice is a basic assumption of the theory of intersexual selection. We have also discussed several reasons why we might expect animals to be selective in their choice of mates. What is important is whether our theories are borne out by the facts. We have seen a number of examples where evidence of mate choice has been obtained, though in some of these it is not clear exactly how choices are made or what benefits animals derive from them. The problem is that mate choice is a very difficult phenomenon to observe. While male sexual behaviour often consists of very dramatic and stereotyped postures, movements or sounds which are easy to observe and describe, females can express a preference for one male rather than another by a barely discernible change in behaviour. When a female approaches one male rather than another we cannot be sure that it is because she prefers that male or whether she was going in that direction anyway. There are many biologists who are intrigued by the theoretical possibility that animals choose their mates and who are responding to the challenge of trying to demonstrate that sexual choice does actually occur in nature. In the years to come we may expect much new data on this subject which may or may not show that our theories are correct.

6   Competing for a mate

The previous chapter looked at the largely passive but selective sexual behaviour typical of females in many species. Our attention in this chapter turns to the sexual strategy of males which, in a great many animals, involves some kind of competition for females. As we saw in Chapter 2, the evolutionary basis of this competition is that males typically have a much greater reproductive potential than females. We also saw that male competition may be expressed in one of two ways. Males may either compete to be more attractive to females than their rivals or they may seek to establish a physical superiority over their competitors through aggression. These two types of competition form the basis of intersexual and intrasexual selection respectively. In the last chapter, we were concerned with intersexual selection, in this chapter we shall turn to the evolutionary results of intrasexual selection, the second of these two kinds of competition between males.

In many species males fight each other, either to gain possession of females directly or to establish ownership of resources to which females are attracted. In many species, intrasexual selection has favoured the evolution of male characters, such as large size and the possession of weapons, that enhance fighting ability. However, as we shall see later in this chapter, fighting is not the only way by which males may compete with each other. In many species, males show extremely subtle patterns of behaviour which serve to enhance their reproductive success at the expense of their rivals and which cannot be called aggression.

Large size, strength and weapons

Full-blooded fighting is a feature of the mating activity of elephant seals. Bulls rush towards one another with as much speed as is possible for an animal so ill-adapted for movement on the land. They then rear up and butt their chests together, at the same time tearing at each other's heads with their teeth. The benefits that fall to the winners of these fights are considerable. The few males that become dominant bulls during a mating season mate with nearly all the cows, while the losers hardly mate at all. Not surprisingly, this variation in male mating success has led to very powerful intrasexual selection in male elephant seals which has favoured the evolution of large body size in bulls, who are three times as heavy as the cows.

The same effect is apparent among other groups of animals such as the primates (apes and monkeys and man) and the ungulates (deer and antelopes). Those species in which matings are most unevenly distributed among males show the greatest differences in male

and female size. Among the primates, there is a marked contrast between the African savannah-dwelling patas monkey *Erythrocebus patas* in which successful males hold harems averaging seven females, and a species such as the South American night monkey *Aotus trivirgatus* which is monogamous. The male patas monkey is nearly twice as heavy as the female while the male night monkey is actually slightly smaller on average than his mate. In the European red deer *Cervus elaphus*, a polygynous ungulate, mature stags are one-and-a-half times as big as fully grown hinds.

Large size is not the only physical manifestation of intrasexual selection shown by red deer stags. Their antlers, which become larger and more elaborate as stags grow older, provide an excellent example of the evolution of male weapons that are used in competition for mates. With their many sharp points, antlers are potentially lethal weapons, though it is comparatively

rare that a fight between stags results in serious injury. This is because, during a fight, two stags lock their antlers together, each effectively neutralizing the danger from his opponent's points. Fights are not usually settled on the basis of which stag inflicts the most wounds but are a contest of overall physical strength. With their antlers locked, fighting stags push against each other, each trying to drive his rival backwards. Fighting also requires skilful footwork in the struggle to gain a position on higher or firmer ground, and the greater experience of older stags probably contributes to their ability usually to defeat younger males. However, despite his strength and experience, a high ranking stag will quickly lose status and his harem of hinds if his antlers are broken. Unable to push properly against an opponent, he is unable to make effective use of his other advantages: his weight and experience.

Red deer have been studied in great

1

2

3

A dispute between two red deer stags on the island of Rhum. After roaring at each other, the two stags start parallel-walking (1 & 2). Neither is prepared to give way. They turn towards each other (3) and lock antlers (4). After a long struggle during which they push against each other and compete for a position on higher ground, one emerges as the victor and chases his beaten rival away (5).

4

5

detail on the island of Rhum off the west coast of Scotland. Fighting between the stags is confined to the mating period, called the rut, which occurs in October. During the rest of the year, stags are tolerant of one another and frequently move about in groups. In April, they shed their antlers, and it is not until late June that these have fully regrown. Each year the antlers regrow, usually with one more point than in the previous year. Stags differ from hinds not only in having antlers but in growing a mane of shaggy hair that covers their enormously powerful neck muscles. As the rut approaches, the larger, older stags start to mark out territories, scent-marking by rubbing scent glands on their heads against foliage and by urinating frequently. They also roar repeatedly. If a territorial stag is approached by a potential rival he displays by roaring and by thrashing the vegetation with his antlers. If the intruder persists, the contest may develop into a behaviour called parallel-walking in which the two stags pace backwards and forwards keeping a distance of about 10 metres between them. Still, the intruder may not be deterred by the physical appearance and roaring of the territory owner but lowers his head and attacks.

Fighting between red deer stags is a serious and costly business. Despite the fact that their method of fighting tends to neutralize the danger of lethally pointed antlers, a number of stags are seriously injured during the course of a rut. They may break a leg or an antler, or lose an eye. Furthermore, their harem of hinds may have been dispersed by other males during the fight. Even if they avoid injury, stags are severely weakened by the time the rut is over and winter has set in. For the past month or so they have spent nearly all their time either running here and there keeping their hinds together, or roaring, or marking their territories, or chasing away and occasionally fighting with other stags. As a result they have

very little time to feed during the rut and by the time mating activity ceases they are thin and weak. Some stags are unable to survive a severe winter. What has a stag gained to make all this effort worthwhile? Rather less than one would expect. A successful stag may hold a harem of 20 or more hinds during the course of a rut. However, this does not mean that he will father as many calves. As many as half of his hinds may be immature and some of the mature ones may not breed in a particular year, so a harem of 20 hinds may contain only about eight that actually produce calves. A third of the calves will die, either shortly after birth or during the course of their first winter, so that a stag is not likely to produce more than about five calves as a result of one breeding season. What is also important is that few stags succeed in holding harems for more than four consecutive years.

Some of the most spectacular examples of male weapons are provided by various species of horned beetles. Beetle horns occur in a variety of shapes and sizes, those of scarab beetles being projections of the hard covering of the head of the thorax. The horns of stag beetles are not true horns at all but greatly elongated mandibles or jaws. In some stag beetles, the head and mandibles are longer than the whole of the rest of the body. A very high proportion of the male's reproductive effort must go into the development of these huge weapons. In the Hercules beetle *Dynastes hercules*, an inhabitant of Venezuelan forests, males have two forward projecting horns one above and one below the mouth, each equipped with teeth-like projections. Two males fighting over a female begin by grappling each other's horns. If one male manages to get a grip with both his horns outside those of his rival, he has gained the upper hand. He then slowly shifts his grip back over the other male's head and thorax until he is gripping his opponent's abdomen. Now in complete control, he

Fighting between two male stag beetles *Lucanus cervus*. The two males grapple with each other's horn-like jaws (top). The male who is able to get his jaws around the body of his rival is then able to carry him away from the female and dash him against the ground (bottom).

rears up until only the tip of his abdomen and his hindlegs are on the ground and holds his hapless opponent aloft before slamming him down savagely on to the ground, often wounding him severely. Sometimes the victor will carry his beaten opponent far away from where the female is before the fight reaches its violent dénouement.

There are serious disadvantages associated with fighting between males for the possession of females. A male may be so weakened as a result of a fight that, even if he wins, he may be unable to reap the benefits of what he has gained. In a number of polygynous species, such as the red deer and elephant seals, it is not uncommon to observe males with serious wounds that inevitably reduce their capacity to fulfil the normal functions of life like feeding and avoiding predators. If a fight is long and severe enough to result in wounding, it is likely that the protagonists are evenly matched for strength and therefore that they have both received injury. It will thus be to the advantage of both animals if they can settle their dispute with a minimum of actual fighting. The winner will gain the contested female and will be in a fit state to mate with her and the loser, though he loses an opportunity to mate, at least survives unscathed and fully fit to contest future disputes effectively. In most animals in which competition between males is intense, physical fighting is therefore rare and usually occurs only after a prolonged series of interactions involving an exchange of displays, which in most instances are sufficient to settle the contest before it escalates into a real fight. As we saw earlier, red deer stags usually begin a contest by roaring at each other. The roars of a powerful stag are often sufficient to deter a weaker intruder from making any further challenge. If they are not, the contest moves into the parallel-walk phase, which again may result in the intruder giving way before the fighting stage is reached.

Signals which shorten fights

Fights between males could often be avoided altogether if each protagonist was somehow able to assess accurately the strength of his opponent without actually having to fight him. In many species, aggressive displays seem to be adapted so as to emphasize the size and weapons of a displaying male, possibly enabling his opponent to assess his likely chances of winning or losing. But it is often suggested that threat displays make animals appear larger than they really are. We would only expect an accurate signalling system to evolve if the information it conveys about a male's strength or fighting ability leads to an outcome that is advantageous to both protagonists. If aggressive displays do not convey reliable information, natural selection would not favour individuals that respond to them as if they did. For a system of accurate information exchange to evolve, it must be of some advantage both to the displayer, as a result of his signal being accurately interpreted, and to the receiver, who is able to make the most appropriate response. In the past, theories about the evolution of aggressive displays have suggested that they serve to deceive the receiver as to the fighting ability of the displayer. However, more recent theories have proposed the opposite view, that natural selection will favour aggressive displays that convey accurate information that enables disputes to be settled quickly, to the benefit of both participants. One species in which an accurate system of communication has evolved in association with fighting is the European common toad *Bufo bufo*.

For a male toad, competition for females is intense. A toad breeding pond contains many more males than females.

While most males remain in the pond for the full two- or three-week mating period, females stay in the pond only for the two or three days that they need to find a spawning site and to lay their eggs. Thus, on any one day, only a fraction of the female breeding population is present in the pond. Males start to come to the pond a few days before the first females arrive, travelling at night across the land from their winter hiding places. However, the largest migrations of males are on those nights when the females are also on the move. The majority of females are intercepted by a male before they reach the pond with its expanding population of waiting males. A male who encounters a female jumps on to her back and clasps her firmly under the arms in a position called amplexus. So encumbered, she continues on her way to the pond. Other males who come close to the pair also try to jump on, and the amplexed male lashes out with his powerful hind-legs in an attempt to keep his rivals away. If the second male manages to get a grip on the female, a prolonged struggle ensues which may go on for several hours and which continues after the female has reached the pond. Once there, other males may climb on, and occasionally a large ball of toads is formed with a dozen or more struggling males totally smothering the unfortunate female. A few females die as a result of these mass struggles, probably from asphyxiation.

Most fights over female toads involve just two males who not only struggle with each other to gain the best grip under the female's arms but who also repeatedly produce soft, peeping calls. These calls are a very important part of the male toad's fighting strategy. The outcome of a fight between two male toads depends on their relative sizes. While small and large males are equally likely to be the first to clasp a female, by the time a female starts to spawn, about a day after her arrival in the pond, most of the small males have been displaced

Most female common toads are clasped by a male before they reach the pond (top). However, the male has to contend with many rivals before he can mate with the female. Another male has climbed on to the back of the female and is trying to push the first male off (bottom).

by larger ones. Large males produce a deep croak, small ones a rather squeaky one. This is because the frequency of the sound that a toad makes is determined by the length of his vocal chords. Just as a long, taut piece of string resonates with a lower frequency than a short one, so long vocal chords produce a deeper sound than short ones. A small male is incapable of producing a deep call, and thereby appearing larger than he really is, because the length of his vocal chords is determined by the size of his body. Thus the call made during fighting can be used by a male toad as an accurate way of assessing his rival's size and thus his chances of being able to displace his rival from the female's back.

The possibility that male toads may settle their fights quickly by using the pitch of their opponent's call as an indicator of his size, has been tested by a simple experiment. A small or large male was placed on the back of a female and prevented from calling by a rubber band passed through his mouth like a horse's bit. A second male, of medium size, was then placed near to the pair and, when he started to attack, a recorded call was played from a loud-speaker placed just above the pair. This recorded call was either of the same or of a different pitch to the call that the defending male would have made had he not been muted. The attacking male was much less likely to persist in his attack if he heard a deep croak than if he heard a high-pitched one, confirming the hypothesis that the pitch of the call is used as an indicator of the opponent's size. However, the pitch of the defender's call is not the only cue used by an attacking male in making a decision as to whether or not he will press home his attack. Males attacked large males who appeared to be making high-pitched calls less vigorously than they did small ones, probably because large males gave them a more powerful kick.

Fights between male toads may last for as long as 12 hours, despite the fact

that they may be shortened by the response of small males to calls of larger ones. Presumably, the longest fights are likely to be those that are between males of similar size. Toad fights are probably the longest observed between animals; in most species fights are settled quickly. Despite their length, toad fights are not damaging to males though, as we have seen, the female may suffer if too many males become involved. Male toads have no teeth or claws with which to inflict wounds on each other. The only costs they may bear through fighting are that they lose opportunities to find other females and that they may become exhausted.

While fights between male toads are partly settled by a cue which bears a direct relationship to competitive ability, disputes in some other species are resolved on the basis of what appear to be quite arbitrary rules. For example, among males of the African hamadryas baboon Papio hamadryas, the outcome of a contest between two males for a female is often not decided on the basis of which male has previously established a position of dominance over the other, but on the basis of which male claimed the female first. The period for which the 'owner' has been with the female may be extremely brief and have involved only the most cursory interactions between the male and the female. Nevertheless, a male of higher rank will often defer to the owner and not attempt to take over his female. There are sound theoretical reasons why we might expect this kind of behaviour to occur under certain conditions. If two animals stand to lose much and gain little if they start to fight, it will be in the best interests of both if they settle their dispute quickly and according to some mutually recognized rule. Baboons risk injury if they embark on a fight and, at the end of it, the female may well have been taken over by another male.

Prior possession is also the rule used by speckled wood butterflies Pararge aegeria in territorial disputes. This

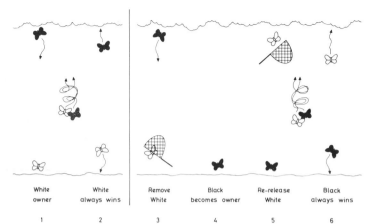

White owner	White always wins	Remove White	Black becomes owner	Re-release White	Black always wins
1	2	3	4	5	6

Above A simple experiment shows that speckled wood butterflies settle disputes on the simple rule, resident always wins. Thus they avoid potentially damaging fights.

Below A speckled wood butterfly basks in a patch of sunlight.

European species lives in woodland, and males compete for patches of sunlight on the forest floor where they court females who come to sun themselves. If a male intrudes into a sunspot the resident approaches him and the two butterflies fly upwards in a brief, spiralling flight before the intruder flies away. It is always the resident who wins these ritualized contests and the intruder who defers. Residence of a sunspot is established quickly. If a male is taken away only for a few minutes, he will defer to the male who has taken his

place during his absence. The new owner may only have been in possession of the territory for a few seconds. As with the baboons, it is a more adaptive strategy for butterflies to settle their disputes in this arbitrary way than it would be to fight. Fighting would be likely to damage their delicate wings and, furthermore, every male is likely to have an opportunity to become the owner of a sunspot for some part of the day as the sun moves across the sky and patches of sunlight shift and change in size.

Fighting between males may be influenced by what females do. As we saw in the previous chapter, female elephant seals may incite males to fight by protesting loudly when a bull attempts to mount them. Because they are more likely to protest when mounted by a subordinate male, this behaviour increases the possibility that they will be mated by one of the largest and most powerful bulls.

Alternative strategies

When several males are competing for a limited number of females it is impossible for them all to win. What are the smaller or weaker males to do if they are unable to acquire a female by fighting? We would not expect them simply to give up. Rather, we would expect natural selection to have favoured the evolution of alternative strategies by which weaker males may achieve some degree of mating success. This is indeed what happens in a number of species in which males compete for females. In the American orb-weaving spider *Nephila clavipes*, several males congregate on a web which has been woven by a female who spends most of her time at its centre. The largest male generally establishes possession of the best position for mating, also at the centre. He gains not only by having almost exclusive sexual access to the female but also by sometimes feeding on prey which she has caught. The smaller males, scattered around the periphery

Variation in male appearance and behaviour. A group of male ruffs at a lek showing both dark residents who hold territories and pale satellites who do not.

of the web, very rarely get a chance to mate with the female. However, the female may occasionally eat one of the males and, if it happens to be the central male, one of the peripheral males is on hand to take over the central position.

The males of the Gila topminnow *Poeciliopsis occidentalis*, a fish of south-west North America, adopt one of two mating strategies. Some males are black, defend territories and court females with elaborate displays. Others are brown, non-territorial and simply rush up to females and, without displaying to them, attempt to copulate. While the largest males are territorial and the smallest are not, intermediate-sized males may belong to either type. The territorial and the non-territorial males make a roughly equal number of mating attempts which suggests that one strategy may be as effective as the other, though it is not known for either strategy what proportion of mating attempts results in successful copulation. Males can rapidly change from one pattern of behaviour to the other. If a black, territorial male is removed, a brown male will become black within a matter of minutes and then take over the vacant territory.

In the case of the ruff *Philomachus*

pugnax, a European bird that forms leks, males also vary in both colour and behaviour. However, they are not able to change either their appearance or their mating strategy. Unlike the Gila top-minnow, there is a genetic basis to ruff appearances. Some males, called residents, are territorial and have black or dark brown plumes behind their heads. Others are non-territorial but move about in the lek and are called satellite males. These have very pale or white plumes. Residents vigorously defend their territories against intrusion by other residents but are more tolerant of the paler satellites. The adaptive value of this tolerance is that, although residents lose some matings to the satellites in their territories, the presence of several males makes a territory more attractive to females.

Satellite behaviour is also shown by males of the North American green tree-frog *Hyla cinerea*. Most males adopt a strategy of calling repeatedly to attract females as they arrive in the breeding pond. About one in seven males do not call but take up a position within about half-a-metre of a calling male and attempt to intercept females as they approach the caller. Satellites are fre-quently successful in obtaining females and a calling male often tries to repel satellite males by using an aggressive call, by butting or by fighting. There is no discrepancy in size between callers and satellites and individual males may be callers one night and satellites the next. One possible explanation for this dual strategy is that calling is an exhausting activity and that satellites are simply taking a rest.

A remarkable adaptation shown by males who have not acquired a mate, is seen in the olive baboon *Papio anubis*. In this species, males compete to establish consort relationships with those females who have come into oestrus (*ie* are sexually receptive). A male who does not have a female may enlist the support of another single male and together they attack a male who is in consort with a female. The consort male is driven away and the male who initiated the attack takes his place. On a later occasion the same two males will again join forces but this time it is the other partner who initiates the takeover and who claims the female.

In species in which animals are long-lived and breed in each of several years, young males may be smaller than their elders and thus unable to compete for females on equal terms. In such species, alternative male mating strategies may be related to age. For example, young red deer stags do not attempt to defend territories or maintain harems but adopt 'sneaky' mating behaviour in which they attempt quick copulations with hinds when their stags are not attending to them. It is unlikely, however, that these mating attempts ever lead to successful fertilization.

Many reptiles continue to grow throughout their lives and, as a result, there is considerable variation in the size of sexually mature males. In the green iguana *Iguana iguana*, a species which has been studied in Panama, males may adopt one of three possible mating stretegies, depending on their size. The oldest and largest males defend small territories. These do not contain any useful resources and serve only as a place where mating occurs. A successful territorial male may have as many as four females in his territory at one time. Medium-sized males do not defend territories but tend to be found around the edges of those held by the larger males. They display to and attempt to mate with females that are moving in and out of the larger males' territories, and may become territorial if one of the larger males dies or is removed. Small males are found within the territories of the large males, who are more tolerant of them than they are of the medium-sized males. The small males attempt to mate with any females that are present within a territory. The distribution of matings among the males is determined by the behaviour of females who show a marked preference for the larger, territorial males. There is little fighting between males, territorial and other aggressive encounters being settled by elaborate displays. The green iguana mating system is best characterized as a lek in which the larger males compete for sites from which they can display to females. As in many lekking species, the females are not synchronized in their breeding activity, so that only some of them are involved in mating at any one time.

In some species, the small, younger males do not show an alternative mating strategy. They either opt out of mating altogether until they have grown large enough to have some success in mating competition, or simply do the best they can in rivalry with their larger elders. Male elephant seals reach sexual maturity two or three years later than females, their early years being devoted to growth rather than reproductive effort. In contrast, male toads come to breeding ponds even when they are still small, and a few do mate successfully, though most of them get displaced by larger males.

Sexual interference

Most of the examples of mating com-

Sexual interference in the American tiger salamander. A male (white) has started to lead a female (black) in a tail-nudging walk. A second male (stippled) intrudes and stimulates the 'white' male to deposit a spermatophore (white arrow) by copying female behaviour. The 'stippled' male then deposits his own spermatophore (stippled arrow) on top of that of the 'white' male. Finally, the female is led on to the spermato-phores, but she can only pick up the sperm of the intruder male. At the same time, the 'white' male, still believing he is leading a female, deposits another spermatophore.

petition between males that we have discussed so far have involved aggression, either in the form of actual fighting or in the form of ritualized displays. In many species rivalry between males is not expressed aggressively but in the form of intensive sexual displays directed towards females. The males that are successful are those that are most attractive to females. One way that a male can improve his mating success in this situation is to reduce the effectiveness of his rival's mating efforts. Behaviour that has this effect is called sexual interference. As we saw in the previous chapter, there are certain insects in which the ability of males to attract females depends on their catching insects which they can offer to females during courtship. In a North American hanging fly, *Bittacus apicalis*, a male who does not have a meal to offer a prospective mate sometimes obtains one by imitating female behaviour in such a way that another, prey-carrying male is induced to give away his insect. The mimic is then able to fly away in search of genuine females. A male who shows this behavioural mimicry probably gains in more than one way. Not only does he improve his own chances of attracting a female while reducing those of a rival, but he does so without incurring the risks that are involved in catching insects by normal means. Males who fly about in search of insect prey are more likely to get caught in a spider's web than those who steal prey from other males.

Some of the most striking examples of sexual interference are provided by certain North American salamanders. Mating in salamanders and newts involves the transfer of sperm from male to female in little sacs called spermatophores. The male deposits a spermatophore on the ground or pond floor and the female then walks over it in such a way that her genital opening or cloaca passes close to it. If her cloaca brushes against it, the spermatophore adheres and is drawn up into her body. In sala-

manders, males do not put down a spermatophore unless they have received some indication that the female is fully receptive, usually in the form of a nudge of the female's head against the base of the male's tail. In the American tiger salamander *Ambystoma tigrinum*, a male may approach a courting pair and move in between the male and the female. The intruding male then imitates the female's normal behaviour, nudging his rival's tail and thereby eliciting the deposition of a spermatophore. He then deposits his own spermatophore on top of that of the first male. The female, following the second male, will thus pick up his sperm and not that of the male she originally responded to, whose spermatophore is hidden.

The potential reduction of reproductive success that males can suffer as a result of sexual interference from their rivals has favoured the evolution of counter-measures which are called sexual defence. In the woodland salamander, *Plethodon jordani*, males actively defend females against other males, and, in another North American species, the tiger salamander, males sometimes push females away from rivals before initiating spermatophore transfer behaviour.

Male competition after mating

In many species, females have opportunities, which they may or may not take, to mate with several males. As a result a male, even though he has mated with a female, may turn out to be the father of only some, and possibly none, of her offspring. In such species, various aspects of male reproductive behaviour appear to have evolved as adaptations to ensure the paternity of the offspring of a female that a male mates with. The Uganda kob *Adenota kob* is an antelope in which mating takes place at a lek. Mating occurs throughout the year at mating grounds which consist of a group of very small territories, each defended by a male. A female, who has become sexually receptive, visits the mating ground for

A male Uganda kob nuzzles a female in a post-coital display which is thought to enhance his reproductive success at the expense of rival males.

about a day during which she may mate with up to 10 males. While she moves freely from one male's territory to another, males do not leave their territories, even to pursue females. Males whose territories are at the centre of a mating ground copulate much more often than those with peripheral territories. However, a male's reproductive success may not necessarily be determined by how often he mates. What is really important is how many females he actually fertilizes. Male kob perform elaborate displays to females who visit their territories, not only before, but also after copulation. During his post-coital display, a male spends a lot of time nuzzling the female around her genitals. It is thought that this stimulates the female to secrete the hormone oxytocin which in turn stimulates her uterus to contract. Uterine contractions would assist the upward passage of the male's sperm towards the unfertilized egg. Thus, it is possible that it is the males with the most effective post-coital displays, rather than those who mate most often, who are successful in reproduction.

A female who has mated several times will carry sperm from a number of different males. However, it is often not the case that each male's sperm has the same chance of fertilizing her eggs. In many insects, the last male to mate with a female fathers the majority of her progeny. Thus, sexual competition between males continues within the female's body even after mating. It is not clear exactly what the mechanisms involved in sperm competition are. Many female insects store sperm in a little sac called the spermatheca which has only one opening into the rest of her genital system, serving as both the entrance and the exit for the sperm. It may simply be that the sperm most recently received blocks this opening so that earlier sperm is less able to get out when the eggs are being fertilized. Whatever the precise mechanisms of sperm competition may be, it has had a profound evolutionary effect on the

mating behaviour of males. If a female mates several times before she starts to fertilize her eggs, it will be to the benefit of a male to mate with her, even though she has mated already, especially if he can somehow ensure that he is the last male to mate with her.

In many invertebrates, males increase the probability that it is their sperm, and not that of their rivals, that fertilizes a female's eggs by attempting to retain possession of her from the moment he first meets her to the time when they mate and she lays her eggs. Capture of the female for a prolonged period before or following mating, is called the passive phase. In the sexual behaviour of the dungfly *Scatophaga stercoraria*, the passive phase is extended for a time after mating. Dungflies mate and lay their eggs on newly deposited cow-pats. Males, who are the first to gather around a fresh cow-pat, outnumber receptive females by four or five to one so that competition between them is intense. When a male sees a female, he approaches her, jumps on to her back and starts to mate with her. Her spermatheca already contains a large amount of sperm left over from her last mating and egg-laying cycle. However, this will be displaced by the sperm of her present partner who, provided he can retain possession of her until she lays her eggs in the cow-pat, will fertilize 90 per cent of her eggs. Following mating, the male withdraws his genitalia but maintains his hold on the female until she has laid her last egg. She then sways from side to side, he releases her and they fly away, he in search of further females, she away from the cow-pat. Throughout mating and the passive phase, the male is constantly attacked by rivals. He tries to fend them off with his middle pair of legs, but if a rival succeeds in gaining hold on the female, a long and fierce struggle ensues in the course of which the unfortunate female may get badly damaged and smeared with cow dung. If a male succeeds in taking over a female he immediately

A male dungfly defends a female after he has mated with her. He holds her with his front legs and, at the same time, uses one of his middle legs to deflect a rival male.

mates with her and, in his turn, attempts to keep hold of her until she has finished egg-laying.

In some species, there is a prolonged passive phase before mating. In a relative of the woodlouse, *Helleria brevicornis*, the male may claim possession of a female several days before she is ready to mate. Males are larger than females and carry them around between their many pairs of legs. Mating can only occur just after the female has shed her skin; it is only then that her genital opening is large enough for copulation to be possible. After copulation the male continues to hold the female for a further two days, ensuring that his sperm fertilizes her eggs. A long pre-mating passive phase also occurs in a related species, the European water-louse, *Asellus aquaticus*. Males compete with each other for females and, during the passive phase, a male has continually to try to resist the attempts of other males to take over his female. Large males tend to win these struggles,

though small males often succeed in maintaining a hold on small females.

In the American giant water-bug, *Abedus herberti*, the female lays her eggs on the wide, flattened back of the male and he then looks after them. Like the female dungfly, a female water-bug may carry sperm left over from a previous mating and so there is a potential risk to the male that eggs that are attached to his back, and which he will look after, are not fathered by him. Male water-bugs counter this risk by copulating frequently both before and during the time when the female is laying her eggs. His sperm have then almost complete precedence over any that the female has received during earlier matings, so that his paternity of the eggs on his back is assured.

The problem posed by females receiving sperm from one or more rivals has been solved in quite a different way by the male of an American damselfly, *Calopteryx maculata*. When he mates with a female he inserts his penis into her

A male Australian giant water-bug, *Diplonychus*, carrying his eggs on his back.

genital opening but, before producing his own sperm, he uses his penis to remove any sperm left by a previous male.

Another male device that ensures exclusive fertilization of a female's eggs is a mating plug. In rats and guinea pigs, a substance in the seminal fluid causes it to coagulate in the female's vagina. This fulfils a dual function. A mating plug firmly and correctly positioned in the female's vagina somehow increases the probability that the sperm will successfully reach the eggs. It will also make it impossible for other males to copulate with her. In a number of snakes, mating plugs are formed from a secretion produced by the male's kidneys. The presence of a plug in a female's cloaca not only physically prevents copulation but also inhibits courtship behaviour in other males. Mating plugs occur in insects too. In the biting midge, *Johannseniella nitida*, the mating plug takes a more bizarre form. Following copulation the female eats the male

but his genitalia remain firmly in position in her genital opening.

Some insects employ a quite different method of claiming exclusive mating access to a female. A female mosquito who has mated once may mate with further males, but only the first male will fertilize her eggs. A substance in the male seminal fluid called matrone is responsible for this. This appears to reduce both the female's receptivity to males and their responsiveness to her. Males will begin to court an already fertilized female but often fail to complete their mating behaviour sequence. A similar mechanism exists in houseflies. After a female has mated once, her receptivity to other males is reduced by a component of seminal fluid. In some butterflies of the American genus *Heliconius*, the male applies an anti-aphrodisiac secretion to the female during mating which has the effect of inhibiting the sexual responsiveness of other males to her.

Competition between males may

find expression even after a female's eggs have been fertilized. If a newly pregnant mouse is housed with an unfamiliar male, his novel odour may cause her to abort her developing embryos which were fathered by a different male. She quickly comes into oestrous again, and the new male can fertilize her sooner than he would have been able to had he had to wait until she had given birth. This is only one functional explanation that has been proposed for the phenomenon of spontaneous abortion in the presence of strange males, referred to as the Bruce effect. Another is that it serves to reduce the birth rate in overcrowded conditions when encounters between strangers will become more frequent.

Sexual competition between males is expressed in some species in the form of infanticide. In some parts of India, the langur monkey, *Presbytis entellus*, lives in groups consisting of several females, their young and a single male. In other localities there are several males in each troop. Single-male troops may be attacked by a male who does not belong to any troop and, if he is successful in ousting the resident male, he may kill the young. This accelerates the oestrous cycles of the females who quickly become sexually receptive and are then fertilized by a new male. It is possible that this rather gruesome manifestation of inter-male competition is not a normal feature of langur behaviour but may be a response to abnormally high population densities in certain parts of India. However, infanticide is a regular feature of the social behaviour of the African lion *Panthera leo*. Lions live in prides consisting of several females and their young, and two or three males. Young males are expelled from the pride before they become sexually mature and a threat to the mature males. From time to time, the males holding a pride will be challenged by other groups of males who were expelled from another pride. If they are successful in displacing the resident males, the new males kill cubs that are still being suckled by their mothers. As with the langurs, this stimulates the females to start a new oestrous cycle, bringing forward the time when they can be fertilized by the new males.

This chapter has looked exclusively at sexual competition between males. Females may also show competitive behaviour towards one another as part of their reproductive effort. In many monogamous birds that form durable pair-bonds, the female often joins the male in a cooperative effort to defend their nest or territory. In a colony of herring gulls, for example, a dispute between neighbouring pairs may involve two simultaneous fights, one between the two males, the other between the two females. In species in which many of the normal sex roles are reversed, the female may become the more aggressive and dominant mating partner. This is particularly apparent in polyandrous species, such as the jacana, in which males carry out all parental duties. Females fight one another vigorously for the possession both of breeding territories and of males. In the three-spined stickleback, competition between females is expressed in the form of a dominance hierarchy in which dominant females gain priority of access to male territories. In general, however, sexual competition between females is insignificant when compared with that which has been such a powerful influence on the evolution of male reproductive behaviour.

We have seen in this chapter that the competitive element of male sexual strategy can take a wide variety of forms from full-blooded fighting to more subtle activities such as imitating female behaviour and secreting anti-aphrodisiacs. We have also seen that inter-male competition may be directed, not only towards other males, but also towards their offspring. Moreover, male rivalry can be expressed at any point in the reproductive process. In many

species, a male's mating success may be settled long before mating occurs, by his success or failure to establish a territory. In lions, competition between males is manifested most strongly at a very late stage, in the killing of the cubs of rival males. In a highly sociable species like the lion, rivalry between males can have a severely disruptive effect on the cohesiveness of the social group. In Chapter 8 we shall return to this point and will look at how the behaviour of social species has evolved so as to minimize the disruptive effects of inter-male competition. First, however, we turn our attention to the activity for which much of the behaviour we have just discussed is often the prelude, the act of mating.

Infanticide in lions. A male who has taken over a pride, has stolen one of the cubs and has killed it.

7 Mating

In the three previous chapters we have looked at patterns of behaviour that form the preliminaries to mating. We have seen how animals locate members of their own species who are of the opposite sex, how they choose which of a number of potential partners they will mate with, and how they compete with rivals for access to mates. In this chapter we turn our attention to the behaviour that occurs in association with the actual mating act. In some species, mating is a brief affair, lasting no more than a few seconds. In others, it is prolonged and involves elaborate patterns of behaviour that form a complex series of interactions between male and female which must be performed accurately and in a precise sequence if gametes are to be transferred reliably from one partner to the other.

There are two major aspects to the behaviour that precedes and accompanies the act of mating, which we may characterize as persuasion and synchronization. Persuasion categorizes activities whose function is to increase the willingness of the partner to participate in mating. In many species, the corollary of eliciting sexual responses from the mate is the suppression of behaviour that is not conducive to successful mating, such as aggression, fear and cannibalism. Synchronization of mating behaviour is especially important in those species in which gametes

are not transferred directly by copulation but have to be passed from one animal to the other indirectly or which have to be brought together outside the body. The more complex the manoeuvres that have to be executed to bring the gametes together, the more elaborate are the mechanisms that ensure that each partner does the right thing at the right moment.

Persuasion

Many of the displays that animals perform immediately before mating serve the function of increasing the partner's sexual motivation until the point at which he or she is willing to mate. Usually it is the male who initially is the more eager partner and who performs displays that stimulate the female. The form of these displays varies considerably from species to species, depending largely on the sensory modality by which the female perceives them. In one genus of fruitfly, *Drosophila*, the male vibrates one or both of his wings to produce a low-pitched buzz which the female detects through her antennae. The pattern of sound produced by the male is specific to the species to which he belongs, and, provided his song corresponds to that which is characteristic of her own species, the female becomes passive and allows the male to lick her genitalia before he mounts her. In most birds pre-copulatory displays are visual and involve the male performing a variety of postures. Some species of snail employ a powerful tactile stimulus during courtship. Snails are hermaphrodites, and both partners show reciprocal mating. Each produces a hard, sharp dart which is forced into their mate's body shortly before mating. Many salamanders and newts use chemical cues to stimulate the female, and males show a variety of techniques for conveying secretions from special skin glands to the female. In many salamanders, the male holds the female and rubs his 'hedonic' glands against her nostrils. In one North American species, the two-lined salamander *Eurycea bislineata*, the male applies his aphrodisiac secretion in a most bizarre way. In the breeding season he develops two special teeth that project forwards from his upper jaw, piercing his upper lip, and which play no part in feeding. While holding the female from above he rubs a gland situated under his chin over her body and then lacerates her skin with these teeth so that his glandular secretions enter her bloodstream. He concentrates his efforts around an area on her flanks where there is a large vein very near the skin.

Newts of the European genus *Triturus* differ from other newts and salamanders in that the male does not hold the female at any time during mating or its preliminaries. Instead, during courtship which takes place in water, he performs a complex display which, in the most widespread species, the smooth newt *Triturus vulgaris*, consists of three distinct movements, each of which stimulates the female through a different sensory modality. The three movements are called wave, whip and fan. In the wave, the male adopts a position with his body aligned across the female's forward field of view and gently waves his tail. She thus receives a full view of his body with its large crest and tail and its conspicuous pattern of dark spots and coloured stripes. The whip is a powerful tactile stimulus in which he

lashes his tail against his body, creating a violent blast of water which is often strong enough to push the female backwards. By contrast, the fan is a delicate and sustained movement in which the male curves his tail against his flank and vibrates its tip. This creates a steady stream of water which, from the male's position in front of and slightly to one side of the female, is directed along the side of his body towards her nose. Fanning probably provides both a tactile stimulus, the stream of water stimulating organs in the female's skin that detect water movements, and an olfactory one, the current carrying the male's odour to her.

The male of a closely related species, the palmate newt *Triturus helveticus*, has evolved a number of anatomical adaptations that are associated with his fanning display. In the breeding season he develops a glandular ridge that extends along each side of his body. This forms the upper edge of a groove that runs along his flank and which is bounded below by the swell of his belly. When he displays to a female he vibrates the end of his tail in the posterior part of this groove which channels the water current over the glands in his skin. The male also develops a short filament at the tip of his tail during the breeding season. This too is probably associated with his fanning display, serving to reduce the turbulence created by the rapidly vibrating tail tip, and thus helping to produce a more even flow of water, fulfilling essentially the same function as the wires that trail from the rear edge of many aircraft wings. The presence of these anatomical features, whose sole function appears to be to increase the effectiveness of one of the male's courtship displays, is an indication of how important such displays are to his reproductive success.

In a very few species it is the female who plays the more active role in the preliminaries to mating. Where this occurs it is usually associated with a reversal of the 'normal' sex roles,

The complex courtship sequence of the European smooth newt *Triturus vulgaris* involves minimal contact between male (black) and female (white), but the female must respond to the male or he will break off in mid-display.

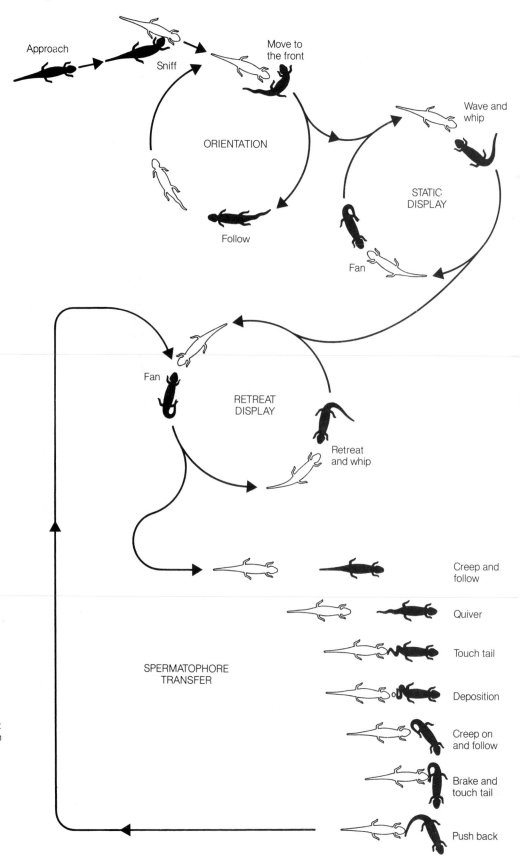

Approach

Sniff

Move to the front

ORIENTATION

Follow

Wave and whip

STATIC DISPLAY

Fan

Fan

RETREAT DISPLAY

Retreat and whip

Creep and follow

Quiver

Touch tail

SPERMATOPHORE TRANSFER

Deposition

Creep on and follow

Brake and touch tail

Push back

Opposite Throughout mating in the Spanish pleurodele newt *Pleurodeles waltl,* the male keeps a tight hold on the female. Locking his arms over hers, he carries her about on his back.

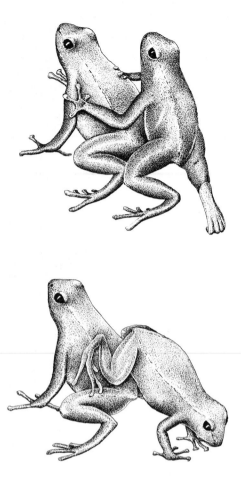

In a rare reversal of 'normal' sex roles, a female poison-arrow frog, *Dendrobates auratus,* solicits mating by stimulating the less ardent male on his back using her fore- and hindlimbs.

whereby it is the male, rather than the female, who carries out most of the parental duties. One such species is the Panamanian poison-arrow frog, *Dendrobates auratus,* in which the male tends the eggs and, when they have hatched, carries the tadpoles to water on his back. While the male initiates contact with the female by calling, it is she who takes the more active part once the pair have come together. She solicits mating by a series of elaborate movements in which she clambers on to the male's back, thereby giving him tactile stimulation. In one movement she performs a hand-stand, with her forelegs on the ground and her hindlegs drumming on the male's back.

Hermophrodites

That one partner is usually more active than the other in the preliminaries to mating is a reflection of the fact that the two sexes allocate their reproductive effort in different ways. Typically, it is the sex that makes the smaller parental effort that is the more active. This disparity does not exist in most hermaphrodite species where both partners in a mating pair are both male and female and allocate their reproductive effort in similar ways. In most hermaphrodites, mating involves the simultaneous transfer of sperm from each participant to the other. Earthworms mate above ground on mild, damp nights, each partner coming halfway out of its burrow. They position themselves side by side, their heads pointing in opposite directions, and secrete a mucous sheath that binds them together, if they are not disturbed, for as long as three or four hours. In this position sperm is exchanged via grooves that run along the surface of each worm's body.

In slugs and snails, courtship culminates in the reciprocal insertion of each partner's penis into the female opening of the other. Several species of slug perform this mutual mating act following remarkably acrobatic preliminaries. A pair first spend an hour or more circling around each other on a tree or wall, producing a mass of mucous slime as they do so. Eventually they twist around each other and drop into space. As they fall, the mass of slime becomes a cord from which they hang, twisting to and fro in mid-air. In this position they insert their penises and exchange sperm before either dropping to the ground or climbing back up their mucous rope.

A quite different pattern of mating is shown by an hermaphrodite fish, the black hamlet *Hypoplectrus nigricans,* an inhabitant of tropical Atlantic waters. Two fish come together and, during a two-hour period just before sunset, take it in turns to play the male and female roles. One fish produces a batch of

eggs into the water and its partner fertilizes them before they are swept away into the plankton. Then the two fish reverse roles, the father of the first batch of eggs producing a batch of his own. This alternation continues with each fish producing an average of four or five batches of eggs, until one or both have laid all their eggs.

Avoiding being attacked

Whereas mating between hermaphrodites involves both partners performing similar and reciprocal patterns of behaviour, in animals in which the sexes are separate, one partner often responds to the sexual advances of the other with behaviour that seems to bear little relationship to mating. In some species, the male runs considerable risks when he attempts to mate with a female who may attack and eat him. Cannibalism associated with mating is a feature of a number of species of insects and spiders. While males in some species have evolved mechanisms that prevent their being eaten, others are not so fortunate. In some mantids, the female eats the male even while he is mating with her. Despite the fact that she begins by devouring his head, the rest of his body continues to function and his sperm are successfully passed into her body. By the time she gets to his abdomen, mating has been completed. His sacrifice is not entirely in vain. The nourishment which she gains by eating her mate enables the female to lay a batch of eggs that contain a plentiful supply of nutrients for the larvae that will develop within them.

Male spiders tend to be considerably smaller than females and are therefore at great risk of being mistaken for prey and eaten. Males of different species show a variety of adaptations against this threat. Some vibrate the female's web with a distinctive rhythm that indicates to her that it is a suitor and not a prey that has landed on her web. In some species the male presents the female with an insect that he has caught and which serves to keep her mouthparts busy while he mates with her. The male tarantula counters the threat from the female's powerful jaws by holding them open with his front pair of legs which carry curved spurs specially adapted for this purpose. In one species of crab spider, *Xysticus cristatus*, the male appears to immobilize the female by tying her down with a network of silk threads. He then creeps under her and injects his sperm into her genital opening. In fact the female is not physically prevented from moving by the tangle of silk threads, whose function is probably to stimulate her and to enable the much smaller male to climb around her.

Courtship feeding is used as a device for suppressing the predatory tendencies of the female by certain species of flies belonging to the family Empidae. In these creatures, the basic mating pattern is that the male gives the female a gift of an insect which he has previously caught, and mates with her while she is busy eating it. However, different species in the empid family show a variety of elaborations and modifications on this basic pattern. In some species, the male wraps the prey in silk before he gives it to the female. Since she has to unwrap it before she can eat it, the male gains extra time in which to complete mating. While some species simply wind a few perfunctory strands of silk around their gift, others enclose it completely in a silk cocoon. There is one species in which this pattern of behaviour has evolved an extra twist. Before wrapping up the prey, the male sucks all the juices out of it, so that all the female finds when she has unravelled the cocoon, and the male has completed mating, is an empty, non-nutritious husk. While most of the empid flies are carnivorous, and therefore pose a threat to their mates, other members of the family have evolved a different method of feeding and suck nectar from flowers. Despite the fact that the male in such species is no longer at risk of being eaten by the female, he still retains the

Aggression and sex. When they first meet, a male and female black-headed gull threaten each other by showing their brown face masks (top). An appeasement gesture in which the face masks are turned away, enables them to get close enough to establish a pair-bond which eventually culminates in mating.

habit of presenting her with a mating gift, though in a much modified form. What he gives the female is a cocoon of silk which is not woven around an insect but is quite empty. What has happened during the evolution of this behaviour is that the presentation of a gift has assumed a quite different biological significance. The male is no longer giving the female food to distract her while he mates. Rather, the presentation of a silk cocoon has become a ritualistic but essential pre-copulatory display. Females will only mate with males that are carrying cocoons.

Cannibalism is only one response that an animal may show to its mate that will tend to interfere with mating. Many animals lead essentially solitary lives for most of the year, and tend to be aggressive when they meet other members of their species with whom they may have frequent disputes over food and other limited resources. Conse-

quently, when they come together for mating, males and females are often extremely wary of one another and their behaviour may be highly ambivalent, showing elements not only of sexual behaviour but also of aggression and fear. This is particularly true of a territorial species, like the stickleback, for whom an intruder into a male's territory may be a rival or a potential mate. If mating is to occur these other responses to the partner must be suppressed. During the preliminaries to mating many animals perform movements called appeasement displays whose effect is to reduce the tendency of the partner to run away from or to attack the displayer. It is a common feature of appeasement postures that they show elements which are the antithesis of elements of aggressive displays. For example, the European black-headed gull *Larus ridibundus* turns its dark brown face mask towards a rival during an aggressive interaction. In

Right The risk run by this tiny male *Argyope* spider is clear. He could easily be mistaken for prey by the much larger female.

Opposite A male East African mantid, *Sphodromantis viridis,* continues to mate with the female despite the fact that she has eaten his head. The nourishment he provides will contribute to the production of their offspring.

The male fly, *Empis livida,* has collected an insect and awaits a chance to mate with a female (left). By giving her a gift, a male distracts the female, and has time to mate with her while she is busy eating (right).

contrast, a male and a female going through the first stages of pair-bond formation frequently turn their heads away from each other so that their face masks are hidden.

The incompatible tendencies of sex, fear and aggression are apparent in the mating behaviour of many animals. When a male European chaffinch *Fringilla coelebs* meets a female during the breeding season, he shows his sexual interest in her by performing a head-forward posture. This is very similar in form to the threat posture that chaffinches perform in many situations, for example in disputes over food. In the sexual context, the male presents a lateral view to the female rather than the more threatening head-on view presented during an aggressive inter-action. This display has been inter-preted as a compromise between con-flicting tendencies in the male to attack the female, as revealed by his head-down posture, and to flee from her, apparent from the fact that he does not face her head-on. Gradually the male moves closer and closer to the female and, as he does so, his display changes from the head-down aggressive posture to a head-up position that is charac-teristic of fearful birds in an aggressive context. Eventually he gets close enough to copulate with her, after which he flies away and gives a call similar to that given by chaffinches when they see a predator, again suggesting that his underlying motivational state contains a large element of fear. Thus it appears that during the early phases of court-ship a male's sexual motivation is in conflict with aggressive responses to-wards the female, but that, as he gets closer to her, his aggression turns to fear.

The mechanisms of fertilization

The nature and complexity of the be-haviour that accompanies the mating act depends very much on the mech-anism by which eggs and sperm are brought together. In many animals, fertilization occurs inside the female's body; the sperm is inserted into her during mating and usually by means of some kind of intromittent organ. By contrast, a number of species, such as frogs and toads, which mate in water, show external fertilization; male and female release their gametes into the water. A third mating technique, in-volving internal fertilization but an indirect method of sperm transfer, is employed by a wide variety of species in which the males lack an intromittent organ.

A male intromittent organ, or penis, has evolved independently in many groups of animals. Some invertebrates are noted for the length of their penises; barnacles have two which are each 40 times as long as their body. Snails have a long, whip-like penis which is extruded from an aperture on the head. Among the vertebrates, the evolution of a penis has been an important element in the progression from aquatic ancestors, the fishes and amphibians, to the fully ter-restrial reptiles and mammals.

Fish do not possess a penis though some, like the guppy, have specially modified fins called gonopodia that are used to insert packets of sperm into the female. However, in the majority of bony fish, from herrings to trout, the eggs are fertilized externally. Herrings come together in shoals to breed and release their eggs and sperm simul-taneously. This is rather haphazard, and to ensure at least some success in fertilization, several million gametes are released by each individual. A male and female brown trout *Salmo trutta* undergo a sequence of actions which prepare them to release their eggs and sperm at the same time. The males of the cartilaginous fishes, which range from the largest sharks to the common dogfish, have a pair of erectile claspers which they insert into the female as each fish coils round the other.

Among the amphibians, most newts and salamanders employ internal ferti-lization, sperm being transferred from

A few of the many thousands of eggs produced by a perch. Large numbers counter the risks faced by each egg left unprotected in a hostile environment.

the male to the female indirectly in a spermatophore. All frogs and toads, with one exception, practise external fertilization. The exception is the tailed frog *Ascaphus truei* of the American northwest. This species has a unique organ that is neither a true tail nor a penis, but an externally extended cloaca which is used during mating to introduce sperm directly into the female's vent. This organ represents an adaptation to the fast-flowing mountain streams in which the tailed frog lives, an environment in which external fertilization would be highly unreliable.

Among the reptiles, at least two different kinds of male intromittent organ have evolved quite independently. One group of reptiles, represented by a single living species, the tuatara lizard of New Zealand, has no penis at all. The turtles and tortoises have a single penis which has a groove along one side that carries the sperm. Crocodiles have a similar organ. Snakes and lizards, however, have twin penises or hemipenes,

though this is really a misnomer since they do not represent half penises. Each hemipenis is a sac which is turned inside out during copulation like the finger of a glove. In many species there are a number of spines on the inside of this sac which thus project outwards during copulation and hold the hemipenis in place inside the female's cloaca. The male uses only one of his hemipenes at a time, depending on which side of him the female happens to be.

Internal fertilization is an essential feature of reproduction in reptiles because they produce eggs with hard, impervious shells. Since the shell has to be put around the egg before it is laid and because the sperm could not penetrate the shell, it is clearly essential that the sperm and the eggs come together inside the female's body. The same applies to birds who have inherited a hard-shelled or cleidoic egg from their reptilian ancestors.

It is therefore rather surprising that most birds do not have a penis, though

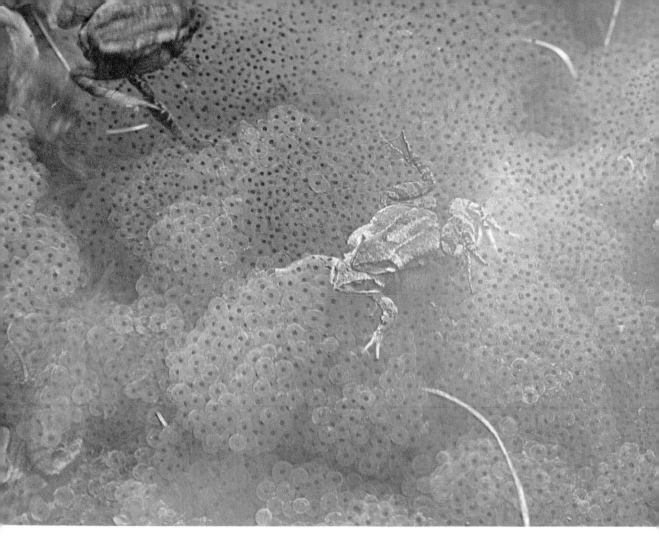

External fertilization. Masses of eggs produced by several pairs of common frogs *Rana temporaria*. The male sheds his sperm as the female expels her eggs. The large numbers of eggs ensure that some will survive.

some have a specially modified cloaca that acts as an intromittent organ. In ducks and geese this consists of a spiral sac along which runs a groove that carries the sperm. In mammals, the penis is an erectile organ which acquires the rigidity necessary for copulation by the hydraulic pressure of blood. In some species, such as badgers, whales and most primates, the penis has additional support in the form of a penis bone, or bacculum. In some species the penis is equipped with hooks that keep it in place during mating. In the cat, the tactile stimulation from these hooks stimulates the female to ovulate.

While the possession of a penis means that the male is bound to be the partner that takes the initiative during copulation, successful mating usually depends on the female showing appropriately cooperative behaviour. In rats and mice, the female must adopt a distinctive posture called lordosis, in which she lifts her rump into the air, before the male can mount her successfully. In tortoises the male has a difficult enough time trying to keep his balance on the female's domed shell, but an unreceptive female can completely foil his efforts by pushing the rear edge of her shell down into the ground. In most species that copulate, it is difficult if not impossible for a male to copulate with a female without her cooperation. Rape, in which the female is very obviously uncooperative, does, however, occur in ducks and does lead to fertilization.

Synchronization
While copulation is often completed very quickly, mating that involves external fertilization is usually associated with prolonged and complex

interactions between male and female. The elaborate courtship patterns of many externally-fertilizing species have evolved largely as an adaptation that ensures that eggs and sperms are released into the outside world at precisely the right moment so that the risks of their being dispersed before they can meet and fuse are minimized. One of the best examples of a complex courtship sequence that ensures precise synchronization of behaviour is provided by the three-spined stickleback *Gasterosteus aculeatus*.

Stickleback courtship consists of a series of interactions between male and female in which each activity is both a specific response to what the partner is doing, and the stimulus that elicits what the partner does next. The sequence thus forms a chain of stimuli and responses. When a female enters a male's territory where he has already built his nest, he responds by performing a zig-zag dance in which he alternately swims away from and towards the female. He continues this display until she responds by turning towards him with her head raised. He then leads her to the nest and, if she has followed him, he adopts a head-down posture near the nest entrance. If she has not followed him, he reverts to his zig-zag dance and the sequence is repeated until she does come down to his nest. Eventually she responds to his head-down display near the nest entrance by entering the nest. He then nudges the base of her tail with his snout, she lays her eggs and leaves the nest through the end opposite the entrance. The male immediately follows her through the nest, shedding his sperm on to the eggs as he does so. In general, neither animal goes from one stage in the sequence to the next until his or her partner has shown the appropriate response. This ensures that both partners reach their peak sexual receptivity, expressed in the production of gametes, at the same time. If one partner is initially more strongly motivated than the other, the fact that he or she does not proceed to the next stage until receiving the appropriate stimulus to do so, allows the less motivated partner to catch up.

It is important that the male stickleback fertilizes the eggs as soon as

The result of internal fertilization. A young gecko has just emerged from its egg where it was protected by the hard shell from desiccation and from, at least, some enemies.

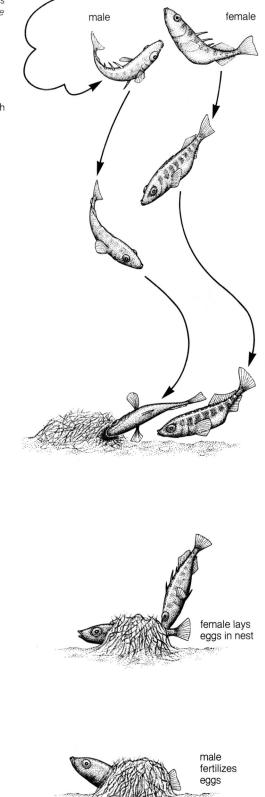

male

female

female lays eggs in nest

male fertilizes eggs

possible after the female has laid them because there is a risk that any delay will provide an opportunity for other males to fertilize them. Male sticklebacks do occasionally approach a courting pair and, given a chance, will rush into the nest as soon as the female has passed through it and fertilize the eggs. As a result, the owner of the nest is left to care for eggs and young many or all of which are not his own offspring.

Fish like the stickleback which build a nest for their eggs thereby provide them with a secure environment in which they can develop relatively safe from the attentions of predators. Other fish employ a quite different technique for guarding their eggs, holding them in their mouths until they hatch. Such fish are known as mouthbrooders, and one example is the African species *Haplochromis burtoni*. This shows a very beautiful adaptation for ensuring that all the eggs are successfully fertilized. The anal fin of both sexes is decorated with a row of orange spots, which are very similar in size to eggs. During courtship the female spreads her anal fin in a display that attracts the male to her. She then deposits her eggs on the bottom of the pond and the male swims over them and fertilizes them. As he does so the female starts to take the eggs into her mouth. Not only does she pick up the real eggs but she also attempts to pick up the egg-like spots on the male's anal fin. As a result she sucks in sperm which is issuing from just behind his anal fin, so that any unfertilized eggs are fertilized inside her mouth. In some species this device has been taken a step further. The male does not fertilize the eggs while they are outside the female's body. Instead she lays them, gathers them up in her mouth and then, while attempting to take up his egg-spots, draws in his sperm instead, so that fertilization occurs entirely within her mouth.

One of the more bizarre courtship procedures associated with external fertilization is shown by the Surinam

toad *Pipa pipa.* In this species, the fertilized eggs are absorbed into the fleshy
skin on the female's back. There they
develop, safely enclosed in tiny pockets,
through a tadpole stage before eventually emerging as tiny toads. A pair of
toads first go into amplexus, the male
clasping the female around her lower
abdomen so that her cloaca is pressed

against his stomach. They then swim
upwards and turn over on to their backs
in mid-water. In this position the female
extrudes a few eggs which fall into folds
on the male's stomach. The pair then
turn the right way up and swim back to
the bottom of the pond. As they descend
the male produces some sperm and the
eggs are fertilized. When they are back

in their original position with the male on top of the female, the fertilized eggs fall out of the folds on his stomach and on to the female's back where they adhere. The pair repeat this procedure several times, fertilizing between three and five eggs on each occasion, until about 50 eggs are eventually attached to the female's back. The male then releases the female, whose skin gradually becomes increasingly thick and spongy until it has completely enclosed the eggs.

As a method of mating, external fertilization has a number of drawbacks. Most important of these is the risk that the eggs and sperms, released into an unpredictable outside world, will be dispersed before they have an opportunity to fuse. Another disadvantage is that mating has to occur at the same time and in the same place as egg-laying, so that opportunities for mating may be very limited. Animals in which fertilization is internal are not restricted in this way, and many such species have taken advantage of the opportunity to emancipate mating from egg-laying. Spiders tend to live at very low population densities with the result that a male and a female are likely to meet only rarely. It is by no means certain that, when they do meet, the time and place will be ideal for egg-laying, or even that the female will be carrying ripe eggs. However, since fertilization is internal, a female can take sperm from a male whenever she meets one, store it in her body, and fertilize her eggs when conditions are most propitious for laying them.

In spiders, sperm is transferred from male to female in a curious, indirect way. The male has two specialized limbs called pedipalps that function as syringes. He first extrudes a drop of seminal fluid from his genital opening on to a specially built web. Then he dips his pedipalps into the drop and fills them up. During mating he applies the pedipalps to the female's genital opening and squirts the sperm into her body.

In some species the male fills his pedipalps before he has encountered a female, in others he does not fill them until he has found a mate and stimulated her into a receptive state. This is just one way of achieving internal fertilization without possessing a penis. Another method of indirect sperm transfer that is used in a variety of animals is to parcel up sperm in little packages called spermatophores. While springtails leave these lying around on vegetation, in other animals the passing of spermatophores from male to female is often the occasion for highly elaborate and complex interactions between mating partners. Some of the most striking and best studied examples of spermatophore-transfer behaviour are provided by the tailed amphibians, the salamanders and newts. Following mating in these animals, there is an interval of days, weeks or months before the female lays her eggs, fertilizing them as she does so. In many species she takes great care to lay the eggs in protected places, something which many animals that employ external fertilization, like frogs and toads, are unable to do.

Internal fertilization has another very important evolutionary significance. If the eggs are fertilized inside the female, there is a potentiality for their being retained inside her body for part of their development. Animals which do this and which give birth to partially mature young are described as viviparous or ovo-viviparous, depending on the extent and nature of the support provided to the young by the mother. When developing eggs are afforded only protection and, usually, respiratory gas exchange, the condition is called ovo-viviparity. This occurs in many fish, amphibians and snakes. In the European fire salamander *Salamandra salamandra*, for example, the female gives birth to large, fully-formed tadpoles whose development inside her was nourished by the yolk in the eggs. When, as in mammals, the developing young receive nourishment directly

A male Corsican brook salamander *Euproctus montanus* captures the female by her tail. He then wraps his tail round her and manoeuvres himself until his cloaca is pressed directly against hers.

from the mother the condition is called viviparity.

Despite the fact that newts and salamanders, collectively called urodeles, are very uniform in the overall features of their anatomy and habits, they show a remarkable variety of behaviour patterns associated with the transfer of spermatophores from male to female. The typical urodele spermatophore consists of a gelatinous base which is secreted by glands in the male's cloaca, on top of which is placed a sperm mass. In the brook salamanders *Euproctus* from Sardinia, Corsica and the Pyrenees, the male captures the female by wrapping his tail around her. In a lengthy process, which may last for two or three days, he manoeuvres himself into a position such that his cloaca is directly pressed against hers so that his spermatophore can be transferred directly. In the European fire salamander, mating occurs on the land. The male captures the female from underneath, holding her by looping his forelegs up and over hers. In this position, he stimulates her by rubbing large glands on the top of his head against the underside of her chin and by undulating his tail from side to side. After some time he produces a spermatophore which adheres to the ground beneath his cloaca. Then he swings his body and tail to one side, out from under the female, while still holding on to her arms. As a result of this movement, the rear part of her body, no longer supported by his, falls on to the ground, her cloaca dropping on to the spermatophore. At no time during courtship in either of these animals does the male release his grip on the female. This is in contrast to other urodeles where we find that, to varying degrees, the female is not restrained by the male.

The red-spotted newt *Notophthalmus viridescens* of eastern North America mates in the water, and the male grasps the female only during the early phases of courtship. He holds her with his hindlegs wrapped around her neck and stimulates her in two ways. Curling his body round, he rubs glands on his cheek against her snout and also wraps his tail around and beats it in such a way

that a stream of water passes forwards over her head. After some time he lets go of the female and climbs off her back, dismounting immediately in front of her nose. Having now released the female, the male crawls away from her with his tail waving and flexed to one side. She follows him and nudges the base of his tail. He responds by stopping and depositing a spermatophore. He then moves away to one side; the female follows him, her nose pressed against the base of his tail, and, as she moves over the spermatophore it becomes attached to her open cloaca.

In the courtship of European newts *Triturus*, which are closely related to *Notophthalmus*, the evolutionary trend towards total freedom of movement on the part of the female has gone a stage further. At no time does the male clasp the female. Instead, as we saw earlier in this chapter, the male stimulates her by means of a number of different displays which he performs with his tail. A female signals her responsive-

ness to a male's display by approaching him. He maintains his display for a while, at the same time retreating before her. He then stops, turns and creeps in front of her for 10 or 20 centimetres. He then stops and quivers his tail which she, following him, touches with her snout. He deposits a spermatophore and then turns to his left or right through a right angle so that his body now blocks her path. His movement following spermatophore deposition is such that he blocks the female at a point where she has moved one body-length from where she was when she touched his tail. As a result, the male stops her when her cloaca is in the vicinity of the spermatophore. If this curious manoeuvre has been carried out accurately the spermatophore attaches itself to the female's cloaca and the sperm are drawn up into her body. The base of the spermatophore consists of a fluid-filled sac. This expands during the four or five seconds that elapse between its deposition and the completion of the male's turning

Spermatophore transfer in the American dusky salamander *Desmognathus ochropheus*. The male deposits a spermatophore and begins to walk away from it (top). The female follows him, keeping her nose pressed against the base of his tail. As he moves away, a mucous thread is stretched from the spermatophore to his cloaca (middle). This may help the female to walk along a line so that her cloaca ends up directly above the spermatophore (bottom).

movement so that the sperm mass is lifted off the floor of the pond thereby increasing the likelihood that it will be picked up. In fact, only about half of all spermatophores are picked up, though during the course of a courtship encounter, which may involve the deposition of two, three or more spermatophores, the probability of a successful transfer increases at each successive attempt.

In another family of urodeles, the plethodontids, the female is likewise not clasped by the male, but the mechanism of spermatophore transfer shows features that are clearly adaptations that reduce the large element of chance that exists in other families. In the North American woodland salamander, *Plethodon jordani*, which mates on land, the female maintains physical contact with the male just before and during spermatophore transfer in a distinctive movement called the tail-straddling walk. After some preliminary interactions the female positions herself astride the

male's tail, with her chin resting on his back at the base of his tail. She maintains this position as the male walks forwards, and he assists her by walking faster if her head comes too far forward or slower if it begins to slip down his tail. After a while he stops and puts down a spermatophore. He then flicks his tail out of the way to one side and continues to walk forward. She follows him, her nose still pressed against his back, until he stops exactly a body-length from the spermatophore. He then presses up with his hindlegs and so pushes up the female's head. At the same time she lowers the posterior part of her body and the spermatophore is met by her cloaca.

In another North American plethodontid, the mountain dusky salamander *Desmognathus ochropheus*, the male produces a very thin mucous thread that stretches from his cloaca to the spermatophore as he moves away from it. As she follows him she keeps this thread just beneath her chin, ensuring that she moves along exactly the right line.

Urodeles are by no means the only animals that transfer sperm in a spermatophore. This device for achieving internal fertilization has arisen many times during the course of evolution, and is also found among insects, arachnids, molluscs and fish. The behaviour patterns that accompany spermatophore transfer in these diverse groups are as varied as the structure of the spermatophores themselves. Of the arachnids, scorpions have perhaps the most elaborate spermatophores. These consist of two parts hinged together; the section that carries the sperm is held down under tension by projections on the spermatophore base. During mating the male locks his two claws with the female's and performs a dance in which he carefully pulls her over the spermatophore. Special hooks close to her cloaca release the hinged portion of the spermatophore which springs up, thrusting the sperm into her body. In some scorpions, the male does not hold the female but instead waves his claws to entice her over the spermatophore. In the cephalopod molluscs, the spermatophore is capable of swimming on its own, having a long worm-like muscular body. That of one octopus species is one metre long. One of the male's arms, called the hectocotylus, is specially modified for passing the spermatophore into the cavity between the female's mantle, a fold of skin that protects the gills and other delicate organs. Once there, it then finds its own way into her genital opening.

We have seen in this chapter that the act of mating, whether it involves copulation, external fertilization or the transfer of a spermatophore, requires that the male and the female act in a co-ordinated way if it is to be successful. The cooperation between the sexes that generally accompanies mating is in marked contrast to the aggression and deception that are often a feature of the behaviour that precedes it. In the next chapter we continue the theme of co-operation in reproductive behaviour. But this time we look not at the behaviour of mates, but at the ways that other individuals may contribute to the successful rearing of young belonging to others.

8 Sex, the family and society

In many species, the rearing of young involves the participation of individuals who are not their parents. In several monkeys and apes, infants are held, groomed and played with by a number of the adult females in their troop besides their mother. Lionesses belonging to the same pride allow one another's cubs to suckle their milk. In many birds a breeding pair are given assistance by others in the defence, feeding and care of their young. These are all examples in which activities normally shown by parents to their own offspring are directed towards the offspring of others. As we saw in Chapter 2, caring for the offspring of others represents one way, along with mating and parental care, in which individuals may expend their reproductive effort. It is this aspect of sexual strategy that forms the focus of the first part of this chapter. We shall see that, for many species, it is not simply the pair, but the family that represents the effective reproductive unit.

Many species live in cohesive social groups which show various forms of cooperative behaviour. Some or all the members of animal societies may be closely related to one another genetically. We have already seen, particularly in Chapter 5, that sex can generate powerful competition and hostility between individuals. In the second part of this chapter we shall examine the interaction between sex, with its tendency to cause social disruption, and other aspects of social behaviour. We shall see that the tensions caused by sex in social species are often mitigated by the close genetic relationships of group members.

Kin selection

Animals that expend part of their reproductive effort in the care of offspring that are not their own pose a major challenge to the evolutionary theorist. Any effort expended by an individual that increases the reproductive success of others is effort that is lost, inasmuch that it is no longer available to be put into that individual's own mating and parental activities. Thus animals that assist in the breeding activities of others are, to some extent, likely to be diminishing their own reproductive potential. Behaviour by an individual that increases the fitness of others while reducing his or her own is defined as altruistic behaviour. Our problem in trying to explain the evolution of altruism is that such behaviour appears not to be explicable in terms of the theory of natural selection. The essence of natural selection is that characteristics are favoured if they confer some benefit on those individuals that express them. The benefit, or increase in fitness, may be an enhanced ability to survive the rigours of the environment or an increase in reproductive success. An activity that is altruistic, that is, which

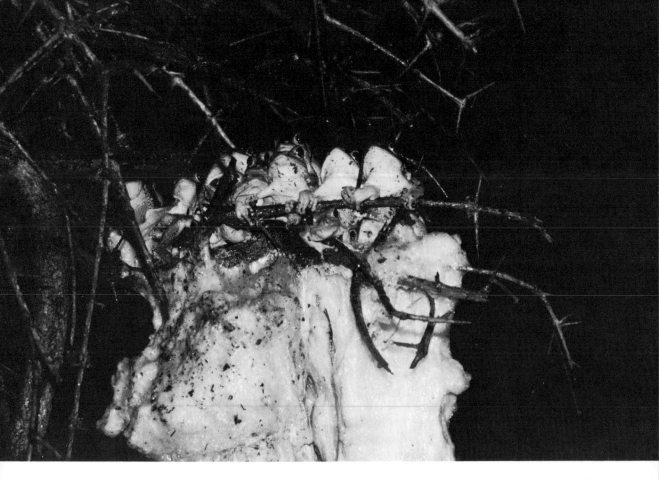

Communal breeding in the frog, *Chiromantis petersi*. A single female produces a liquid which she and several males beat into a foam. She lays her eggs in the nest which the adults then leave. When the tadpoles have hatched, the first rains dissolve the hardened foam, and the tadpoles fall out into the pool beneath. A limited number of water sites may result in large numbers of males and females building huge nests like the one shown here.

reduces the fitness of an individual that performs it, surely cannot be favoured by natural selection?

In the past an answer that has frequently been given as a solution to this problem is that altruistic acts have evolved because they benefit the social group or, more generally, the species. Such a statement invokes the theory of group selection. We saw in Chapter 1 that group selection is entertained by some theorists as an explanation for the evolution of sex, but in the context of the evolution of altruism and other aspects of social behaviour, it is now generally regarded as an unacceptable theory. The reasons for this are rather complex and, for our present purposes, a very brief outline of them must suffice.

Suppose that an individual who behaves altruistically has appeared in a population of non-altruistic animals. Unless the species concerned has, like ourselves, an elaborate system of culture, such a behavioural trait will only spread through the population if it has

some genetic basis. However, the altruistic individual, because of the very nature of his particular attribute, is less likely to pass his genes on to succeeding generations through his own offspring than other individuals who are not altruistic. Altruism can only evolve in a population if all members of that population share the genes that promote it. It is only under exceptional circumstances, such as when an entire population is descended from a single individual and when there is very limited interbreeding between populations, that such genetic homogeneity is likely to occur. However, a more important point is that, even if genes for altruism have become established in a population, any individual who is less altruistic than the rest will tend to leave more offspring. Thus, even if altruism evolves under special and exceptional circumstances, natural selection will not favour its continuation in the face of any occurrence of less altruistic individuals.

What this argument seems to be

saying is that altruism cannot evolve through natural selection at all. How then are we to explain the evolution of the many occurrences of apparently altruistic behaviour in nature? In the great majority of cases that have been studied in detail it has been found that the beneficiaries of an altruistic action are genetically related to the individual that performs it. This gives us the key to understanding the evolution of most forms of altruistic behaviour and, more specifically, the examples of cooperative breeding that we shall discuss in this chapter. Two individuals who are closely related to each other are more likely to have specific genes in common than two unrelated individuals. An individual who carries a gene that promotes altruistic behaviour is likely to share that gene with his parents and his siblings. Therefore, if that individual behaves in a way that increases the reproductive success of one of those close relatives, he is, indirectly, promoting the spread of that gene. This argument, which assumes that selection acts upon groups of close relatives rather than on individuals, is called kin selection.

The evolutionary results of kin selection are most vividly shown in the social insects such as bees and termites. Each individual in a hive of bees plays a specific role in the cooperative reproductive effort, such as foraging for food, guarding the hive, or feeding the developing larvae. The vast majority of bees in a hive are sterile workers who cannot themselves reproduce but who contribute to the parental care of the eggs and larvae, who are all produced by one fertile female, the queen. The important point is that all the members of a hive are very closely related so that sterile individuals are contributing to the proliferation of their own genes even though they cannot reproduce themselves. A bee swarm is essentially a very large, close-knit family.

The theory of kin selection is really nothing more than an extension of the theory of natural selection. It is of no surprise to us that selection has favoured the evolution of the most obvious and widespread form of altruism, parental care. It is self-evident that individuals that behave in ways that enhance the survival of their offspring, even at their own expense, will tend to leave more offspring than those that do not. Sexually reproducing animals share, on average, the same number of genes with their siblings as they do with their progeny. There is therefore no reason why selection should not favour behaviour that benefits siblings and other close relatives just as it favours behaviour that benefits progeny. One species in which individuals assist in the parental care of their siblings, rather than of their offspring, is the scrub jay of North America.

Cooperative breeding

The scrub jay *Aphelocoma coerulescens* is primarily a species of the western United States, but detailed studies of its breeding biology have been made on a small population in the Florida peninsula. In this area, there is rather little of the oak scrub that provides its breeding habitat. Consequently, it is not possible for all birds in the population to find suitable breeding sites, and it is this fact which partly determines their social behaviour during the breeding season. Scrub jays live in groups whose membership is fairly stable and which defend territories against other groups. A group consists of a pair of breeding adults together with their young from one or more previous years. The pair build a nest within the group territory. The breeding female alone incubates the eggs and, while she does so, her mate feeds her. It is when the eggs have hatched that the rest of the group, who are all siblings of the new brood, start to give assistance to the breeding pair. They bring food to the nestlings, often giving it to their mother to pass on to them, and also play a part in warning the rest of the group about nearby predators and in collectively mobbing them.

Just over half of all scrub jay groups in Florida include non-breeding helpers. On average there are two helpers at each nest, though some groups contain as many as six. It is quite clear from the relative fledging rates at nests with and without helpers that they do enhance the reproductive success of the breeding pairs. At nests where a breeding pair are raising their first clutch together, nearly twice as many young are fledged at those where there are helpers than at those where there are none. At nests where the pair have had previous experience of breeding together, the influence of helpers is again significant but rather less dramatic. This enhancement effect is largely self-perpetuating. The more young there are that are successfully reared one year, the more helpers there will be to assist in the next year's breeding effort. The helpers contribute to the breeding success of their group in a number of ways. As well as contributing to the group's collective ability to detect and deter would-be predators, they provide a substantial proportion of the food that is fed to the nestlings. On average the helpers supply about a third of the food, but this does not mean that chicks in nests where there are helpers get more food than those in nests that are tended only by a breeding pair. All chicks receive approximately the same amount of food, but in nests with helpers the parents have to bear less of the burden of providing it. As a result, breeding pairs that have helpers are less severely stressed than those that rear their young unaided. Breeding birds without helpers suffer an annual mortality rate of 20 per cent, while that for those with helpers is only 13 per cent.

It is thus quite clear that breeding birds gain considerably from having their previous young as helpers. What benefits are there for the helpers? An important point here is that all birds will be both helpers and breeders during their lives. While nearly all one year-old birds are helpers, most females will have attained breeding status by the age of three and most males by the age of five. The question we must thus answer is, why do young birds act as helpers rather than taking no part in breeding but conserving their energy until they are old enough to breed? There are several factors that probably favour adoption of the helper strategy when young. For a young male the most important factor seems to be that, to achieve breeding status, he must first be a member of a breeding group. Within each group there is a dominance hierarchy with the breeding male at the top, helping males next and females occupying the lower ranks. Within each sex, older birds are dominant over younger ones. As a group increases in numbers, so its territory expands until it is big enough to be subdivided into two territories. The newly formed territory is claimed by the oldest, highest-ranked male helper. An alternative way for a male helper to become a breeding bird is to take over from a breeding male that has died; again, it is the highest-ranked helper who assumes breeding status. In both situations the attainment of the position of breeding male in a group depends on the prior establishment of high rank among the non-breeding helpers. Another reason why it is advantageous for a young male to be a helper is that it contributes to the group's breeding success, and thus its size, so that when he becomes the breeding male, he will be supported in his turn by a large group and will accordingly have high reproductive success. Finally, the fact that a helper is contributing to the survival of his own kin means that he is enhancing the proliferation of genes for helping which he and they have in common.

For female scrub jays the options are slightly different. They do not breed with males that were raised in the same group as themselves. Were they to do so, they would be mating incestuously with their brothers, half-brothers or other close relatives. Instead, they leave their family group whenever the position of breeding female falls vacant in another

The extended family of the moorhen. One of the two small moorhen chicks is being fed by a larger sibling. The other two large birds are the parents.

group, and compete with other females for that position. If they are unsuccessful, they return to their original group until another opportunity arises.

The helpers in a group of scrub jays do not all make equal contributions to the shared breeding effort. While the oldest and highest-ranked of the male helpers makes as many, or even more, feeding visits to the nest than the breeding male, one-year-old males contribute hardly any food. Likewise, the females, whatever their age, make an insignificant contribution. It thus appears that the effort put into the group breeding effort by a helper is directly related to his or her prospects of becoming a breeding bird in that group. This further suggests that being a helper is somehow a necessary pre-condition of achieving breeding status.

The pattern of helping shown by the Florida scrub jay is similar to that shown by several other species of birds. In the majority of species studied, helpers are the young or siblings of the breeding birds and themselves become breeders later in life. In the European moorhen *Gallinula chloropus* for example, a breeding pair may rear two broods in a year and the offspring in the first brood help to rear the second. It is often suggested that one of the benefits that birds derive from being helpers when young is that they gain useful experience in the various

skills involved in the rearing of offspring and that this experience enhances their own reproductive success when they become breeders.

In most species in which cooperative breeding occurs, there is a clear distinction between breeding birds that have a parental stake in the communal reproductive effort and helpers who do not. A more complex situation exists in the Tasmanian native hen *Tribonyx mortierii*, a large flightless member of the rail family. Within a native hen population, males outnumber females by three to two so that there are simply not enough females for each male to have an exclusive mate. While some birds do form pairs, many live in breeding groups that are trios or quartets, consisting of one female and two or three males. All members of such groups take part in defending the group territory, incubating the eggs and caring for the young. While one male is clearly dominant over his male partner or partners, he does not prevent them from mating with the female. In a trio, the subordinate or auxiliary male copulates with the female about half as often as the dominant male. It is not known whether each male's sperm is equally likely to fertilize the female's eggs, but if we assume that it is, we are faced with an interesting question. Why should a dominant male tolerate the presence of a subordinate who fathers young that he could father himself were he to exclude the subordinate? Two factors are important in answering this question. First, the presence of one or more auxiliary males makes a significant contribution to a female's breeding success that can be sufficient to offset the loss of fitness the dominant male suffers through not being the father of all her progeny. The second and crucial point is that, where two or more males share a female, they are brothers, so that those young which a male has not fathered himself are nevertheless closely related to him.

Cooperative breeding groups, like those of the scrub jay and the Tasmanian native hen, are essentially extended families with a single breeding female providing the focal point. A few species, however, have colonial breeding systems that involve several females. One such species is a Central American member of the cuckoo family, the groove-billed ani *Crotophaga sulcirostris*. A breeding group consists of between one and four monogamous pairs who cooperatively defend a breeding territory and who all lay their eggs in a communal nest where they are incubated, to varying extents, by all members of the group. However, the appearance of harmonious cooperation presented by a group of anis is very deceptive. Within the group sexual competition is intense between females and is expressed in the destruction of eggs by ejecting them from the nest. The females in a breeding group show a dominance hierarchy. The most subordinate female is the first to start laying her eggs, visiting the nest at intervals of one or more days, each time adding another egg. After a time the next female in the hierarchy also starts to lay. At this time the dominant female does not lay any eggs but does make periodic visits to the nest, during which she frequently tosses out one of the eggs laid by her subordinates. Eventually she also starts to lay and at the same time stops throwing out eggs, thus ensuring that she does not eject any of her own. Once all the eggs are laid, incubation begins, with all members of the group, male and female, making a contribution.

The effect of the dominant female's destruction of her subordinates' eggs is to reduce the overall size of the clutch that will eventually be incubated by the group. If a clutch is too big, it cannot be incubated efficiently and some of the eggs will not develop normally. By throwing out some of the eggs of the other females, the dominant bird ensures that the clutch will consist largely of her eggs and will also be of a reasonable size.

The pattern of egg-laying and egg-ejection shown by the females in an ani breeding group means that a dominant bird will tend to have more eggs in the communal clutch than her subordinates by the time incubation is under way. None of the eggs laid by the dominant female are thrown out. However, subordinate females show a number of features that tend to offset this disparity. First, they tend to lay more eggs than the dominant female, thereby compensating to some extent for those that she throws out. Second, they often keep one egg back which they do not lay until after the dominant has started laying her own eggs and ceased ejecting those of her subordinates. Third, they tend to leave longer intervals between successive eggs than does the dominant female, thus reducing the opportunities for her to eject their eggs. In one important respect the subordinate females have a measure of control over the communal breeding system; it is they that initiate incubation. Any egg that is laid after incubation has begun will eventually hatch later than the rest of the clutch. Consequently, the chick that emerges from it will be younger and smaller than the rest and, therefore, very unlikely to survive. Thus, the subordinate females, by starting incubation, effectively call a halt to egg-laying and so force the dominant female to begin laying her eggs and, at the same time, stop ejecting theirs, at a relatively early stage in the breeding cycle. Despite these aspects of the behaviour of subordinate females, the result of all these complex interactions is that the dominant female may have twice as many eggs in the communal nest as a subordinate bird.

The distribution of incubation effort among the various members of the ani breeding group shows very interesting variation. Males, who show a dominance hierarchy that parallels that of the females, participate in incubation to an extent proportional to their status. Thus the male of the dominant pair, who together have the largest number of eggs in the nest, spends more time incubating than any of the subordinate males.

Females, curiously, show the opposite effect. The dominant female contributes least to the communal incubation effort.

In view of the fact that subordinate birds lose a large proportion of their eggs through the destructive activities of the dominant female, it seems curious that they should join breeding groups at all. It would seem that they would do better to breed as pairs, as indeed some birds do. It may be that there is a shortage of nest sites or that some advantage derived from breeding in a group is sufficient to offset the loss of reproductive potential that it entails. It may be that a female, before she can become the dominant bird in a group and thus have a maximum reproductive success for a minimum of effort, must first be a subordinate member of a social breeding group.

The communal breeding behaviour of the groove-billed ani provides a vivid example of an important factor in the dynamics of the social behaviour of many animals. What appears to be a group of animals acting cooperatively towards a mutually advantageous goal is in fact a collection of individuals, each pursuing a slightly different, self-interested strategy. However cohesive and harmonious a social group may appear to be, beneath the surface there are major conflicts of interest between group members. Probably the most potent source of such conflicts is sex. In the remainder of this chapter we shall look briefly at the interaction between factors that promote cohesive social behaviour and the often socially-disruptive influence of sex in species that typically live in stable social groups.

Sexual conflict in social groups

To understand the evolution of social behaviour we have to consider the benefits that each individual member of a social group gains by belonging to that group, rather than by being solitary. There are many different advantages that may accrue to animals that live in groups, but two of the most important

are increased efficiency in feeding and more effective defence against predators. For many species individuals find or handle food more effectively or more quickly if they feed in association with others than if they feed alone. A predator is often both more quickly detected and more effectively repelled by a group of animals than by a solitary individual. However, animals living in a group also incur a number of costs as a result. The most general cost is that living in constant close proximity to others increases the frequency with which occasions arise when individuals find themselves in competition over an item of food, a roosting place, a mate, or any other limited resource. Within limits, the efficiency with which a group of animals finds food is likely to increase as the number of animals in the group increases. However, the intensity of competition for the food that is found will increase in parallel. For any species, living in a certain environment, there will be an optimum group size that represents a balance between the advantages and disadvantages of living in groups of increasing size. Living in a group is not simply a matter of finding food and avoiding predators; at some time during the year sex becomes an additional factor.

As we have seen throughout this book, and particularly in Chapter 5, sex is the cause of considerable conflict between individuals, especially between males. How is a cohesive, mutually advantageous group structure to survive without being torn apart by the effects of sexual competition? In many species it does not. In birds, such as the European great tit *Parus major*, the onset of the breeding season coincides with a complete switch in behaviour from a winter habit of feeding in flocks to a highly aggressive pattern of large breeding territories. Less extreme changes occur in some group-living primates. For the sifaka *Propithecus verreauxi*, a lemur living in Madagascar, for example, the onset of the breeding season is a time of considerable social upheaval. Long-established dominance and subordinance relationships may change, many males leave the troop they have belonged to for many months and join another, and troops generally become rather ill-defined. One of the most detailed studies of the influence of sexual conflict on the dynamics of social behaviour is that of the lion in the Serengeti region of Tanzania.

The single most important factor that binds a pride of lions together is food. Lions are unique among the members of the cat family in that they live and hunt socially. Lacking the speed that enables a cheetah to outpace its prey, or the stealth and agility that enables a leopard to surprise its victims, lions rely on team work. While some members of the pride lie await in cover, others reveal themselves to their prey, driving them into an ambush. This method of hunting enables lions to kill very large animals like wildebeeste and zebra which are more than adequate to provide a meal that will sustain the entire pride for two or three days.

A pride of lions consists of between two and twelve females, their cubs and usually two or three males. The genetic relationships between pride members are important in understanding lion social behaviour. Females remain within the same pride all their lives and are closely related to each other. Some are sisters, others first cousins. The adult males all come from another pride, from which they were expelled before reaching sexual maturity. They are, on average, as closely related as half-brothers and are quite unrelated to the lionesses in their pride.

Males compete for the possession of a pride, which is a necessary condition if they are ever going to breed. Following their expulsion from the pride in which they were born, males form bachelor groups until they are fully mature. They then form groups of two or three, often all born in the same pride,

and together attempt to expel a group of males that already holds a pride. If they are successful they quickly kill any cubs in the newly-acquired pride that are still suckling from their mothers. These cubs are quite unrelated to the new males and represent a loss of reproductive potential since females who are producing milk do not come into oestrus. Having lost their cubs, the females stop lactating and come on heat and the new males then mate with them. Thus the sexual competition that exists between the unrelated present and past owners of a pride is expressed as infanticide. There is also, however, potential reproductive competition between the related males who together own a pride. They do not, however, fight each other or compete in any way for sexual access to females that come into oestrus. This is largely because of a number of remarkable adaptations shown by lionesses that serve to reduce inter-male rivalry.

Female lions do not come into oestrus at any particular time of the year. They do, however, tend to come on heat at the same time as other females in their pride. One effect of this is that each male usually has a female to occupy his sexual attentions, whenever there is an opportunity to mate. If only one female came on heat at a time, the males would tend to compete to mate with her. The effect of female oestrus synchrony is enhanced by the rather unusual inefficiency of copulation in lions. When a female comes into oestrus a male stays with her for two to four days, during which time he copulates with her approximately every fifteen minutes, day and night. Despite this intense sexual activity, there is only a twenty per cent chance that a female will conceive during an oestrus period. Since mating with a female is so unlikely to produce a benefit in terms of young, it is not worthwhile for males to compete with one another for the right to do so. Fighting between males is potentially very costly. Apart from the danger to a

A battered and old male ousted from his pride, spends his final days in solitude.

male's own future survival if he is wounded himself, it is not in his interests to wound one of his fellows, since it would reduce the ability of the males as a group to repel the attacks of rival groups who attempt to take over their pride. Apart from these considerations, there is the fact that the males in a pride are related to one another. If a male allows his close relative to mate with a female, he is indirectly assisting the propagation of genes which he and his relative have in common. The absence of competition between males is shown by the fact that, on some occasions, a male will stop mating with an oestrous female and will allow one of his fellow males to take over.

There seems to be no other functional explanation for why female lions are so unlikely to conceive, despite the persistent mating shown by males, other than that it is an adaptation that has the effect of reducing inter-male competition which would tend to disrupt the social stability of the pride. It is in the interests of the female to minimize male rivalry because any reduction in the ability of the group of males to retain their ownership of the pride may result in its being taken over by another group of males whose first act will be to slaughter the smaller cubs, thereby reducing the females' reproductive potential.

The close genetic relationships between lionesses in a pride explains an interesting feature of their parental care. A female who is lactating will allow another female's cubs to suckle from her. She thus expends some of her parental effort in the care of cubs that are genetically related to her.

The reproductive behaviour of lions provides a beautiful example of how close genetic affinities between the members of a social group provide the evolutionary basis for a number of mechanisms that reduce the potentially disruptive influence of sex. While females are unable to prevent the infanticide that results when one group of males takes over a pride from another group, the sexual competition that potentially exists between the males within a pride is eliminated, partly by the close kinship of those males and partly by various aspects of the females' reproductive physiology.

The lion is not the only species in which aspects of female sexual behaviour are largely responsible for the maintenance of cohesion in a social group. Quite different mechanisms are shown by female acorn woodpeckers *Melanerpes formicivorus*, a species that lives in western North America. In parts of its range this bird shows highly gregarious social habits that are centered around food. Throughout the year, groups of up to 12 adults defend a territory within which they search cooperatively for food. An important part of their diet consists of acorns and nuts which they collect when abundant and store in special storage trees. These have many holes, drilled in them by the woodpeckers, for keeping the acorns and nuts until the winter when they provide nourishment for the group at times of food shortage. In addition, the group breeds communally. The females all lay their eggs in a communal nest and most members of the group share in the incubation of the eggs and the feeding of the nestlings. The participation of most of the group in parental care is largely the

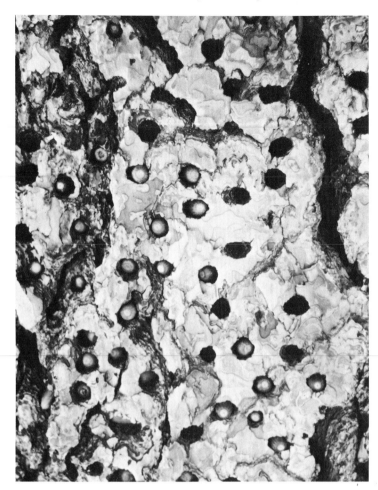

Part of the communal larder of acorn woodpeckers. Numerous holes have been drilled in the bark of a tree, and some have been stuffed with acorns and other seeds which provide the group with nourishment at times of food shortage.

result of two aspects of female sexual behaviour. First, individual females have been observed to mate in quick succession with more than one male, so that one or more males may be the father of their eggs. Secondly, the females all lay their eggs in the communal nest at much the same time so that the parentage of each egg becomes totally obscured. There is no way that a male can be sure which, if any, of the eggs in the nest are his. Consequently, any parental effort that he makes must be directed indiscriminately towards the care of all the eggs and young.

The hypothesis that uncertainty of paternity is the mechanism that ensures that males participate in the communal breeding effort is supported by the behaviour of males who join a group after mating has finished. These males take no part in incubation or in feeding the chicks, though they do participate in communal foraging and food storage activities. This is exactly what we would expect. A male joining a group after mating has finished cannot be the father of any of the eggs in the communal nest and could not therefore gain any reproductive benefit from joining in the communal breeding effort.

An interesting feature of acorn woodpeckers is that they do not always breed cooperatively. In some areas they form monogamous pairs. This enables us to compare the relative benefits of communal and monogamous breeding systems. Groups of three or more birds generally succeed in rearing more chicks than pairs. It is therefore usually more advantageous for a male to join a group than remain in a monogamous pair, despite the fact that he is less certain of the paternity of the eggs and chicks on which he expends his parental efforts in a group.

The social systems we have discussed in this chapter are generally based on long-term relationships between individual animals. The two most important categories of social relationship are those between dominant and subordinate individuals and those between close relatives. Relationships like these can only occur in situations in which individuals are able to recognize one another individually. Recognition of individuals will generally only be possible in social groups that are fairly small and reasonably stable, and in species which have the sensory and cognitive abilities to differentiate between individuals and modify their behaviour in response to previous experience. It is not surprising, therefore, that all our examples have been taken from the 'higher' vertebrates, the birds and mammals, which meet these conditions more often than other groups.

9 Human sexual strategy

The biologist looks at human behaviour in a fundamentally different way from other students of man such as anthropologists and psychologists. He or she tends to adopt a viewpoint from which man is essentially just another species whose behaviour, like that of other animals, has been subject to the influence of natural selection. They sometimes therefore find themselves in the centre of controversy when they start to speculate about the evolution of human behaviour. Some people object to a biological approach to human behaviour on the grounds that it debases man to the level of other animals. Others are wary of a line of argument that says that if we evolved to behave in a certain way, then that is the 'natural' and correct way for us to behave. Another objection is that an evolutionary explanation for a pattern of behaviour carries the implication that it is genetically determined which, to many people, further implies that it is fixed and immutable and therefore not subject to free will or rational thought. While it is undoubtedly the case that evolutionary theories of human behaviour have been used inappropriately, even dishonestly, in the support of political arguments, it is equally misguided to ignore natural selection as an important influence in the shaping of human behaviour.

Some of the objections raised to a biological approach to human behaviour may arise from the way that biologists often study the manifestations of natural selection in the behaviour of animals. We tend to concentrate on examples in which selection has produced adaptations that are beautifully precise and specific to a particular environmental contingency. Particular patterns of behaviour are interpreted as

Human polygyny. A Bahreini man and some of his wives.

A young Danakil
woman from
Ethiopia.

special adaptations to meet particular problems. We have seen many examples of this in earlier chapters. The fact that a row of spots on a fish's tail look exactly like her eggs makes them a vital part of the mechanism by which the eggs are successfully fertilized. There is a serious danger that this concentration on those elements of behaviour whose function is particularly clear and specific leads us to seek simplistic functional explanations for all aspects of behaviour. However, an animal's behavioural repertoire does not consist simply of a predetermined set of specific responses to specific environmental situations. Learning, the ability to alter behaviour in the face of changed circumstances and the invention of entirely new patterns of behaviour are all important aspects of most animals' ability to survive and reproduce. Natural selection does not simply favour behaviour patterns that meet predictable problems; it also favours the ability of an animal to adapt its behaviour in the face of a variable and largely unpredictable environment.

No species illustrates this more vividly than man. In many respects we are a remarkably unspecialized species. Our teeth and digestive system enable us to eat almost any animal and plant food, our limbs give us the ability to walk, run, climb and swim with facility, our hands and brains can turn almost any material in our biological and physical environment into something useful. The variability of human behaviour is very apparent in our sexual activities. Different human societies show enormous variation in their customs and attitudes in relation to sex and family life. While some societies are monogamous, many more are polygynous and a few show elements of polyandry. The exposure of certain parts of the body may be normal in one society and quite unacceptable in another. This variety is reflected not only geographically in differences between societies in different parts of the world, but also in changes that occur over time. In the

west, sexual attitudes have changed to a remarkable degree, going through periods of extreme prudery and of permissiveness within a very short time. Faced with such variety, we are not likely to be able to explain particular aspects of our sexual behaviour in terms of adaptation to clearly defined aspects of our environment. What then can we say about the evolution of human sexual behaviour?

One approach to this problem is to seek features of behaviour that are common to all human cultures. Such features are called 'universals'. One universal in human societies is the existence of some kind of pair-bond or alliance between a man and a woman which shows four important features: some degree of mutual obligation between partners, persistence of the relationship over a prolonged period of time, sanctioning of the alliance by the rest of society, and legitimizing of the offspring. Such a pair-bond is by no means always exclusive; in many societies an individual may form an alliance with more than one partner. Man is unusual among mammals in forming prolonged mating alliances, particularly ones that extend long after mating, through the period of gestation and lactation and beyond. Gestation, giving birth and rearing young are demanding activities for the human mother particularly in primitive environments, and it is unlikely that females would have been able to fulfil them without a male's support. While a prolonged pair-bond is a universal feature of human societies, the mating system of which it forms a part shows great variation from one society to another.

The fact that some aspect of behaviour, such as the formation of sexual alliances, is a universal feature of human societies, cannot be taken as firm evidence that it evolved through natural selection in some primitive ancestor of all the contemporary races. It may be that it arose as a cultural invention of early humans which has per-

In Eritrea, girls and women of the Rashid nomads are veiled even when very young.

sisted throughout the proliferation of human social systems. Alternatively, it may have arisen independently in every society. We cannot assume that we have an innate, immutable propensity to form pair-bonds.

Of some 850 human societies that have been analysed, only 16 per cent practise a monogamous system like that found in western culture. In the great majority, over 80 per cent, polygyny is either the usual or, at the very least, an occasional and acceptable pattern of sexual alliance. In a tiny minority of less than one per cent of human societies, there is some element of polyandry in which a woman forms sexual alliances with more than one man. Examples are provided by certain tribes in Nigeria, the Pahari people of north India and a particular social class, the Tre-ba, in Tibetan society. However, in all cases men may also form multiple alliances. Are we to conclude from the prevalence of polygyny in human societies that man has evolved to be a naturally polygynous species, and that monogamy is therefore unnatural? Such a conclusion ignores an important feature of human mating systems which is that they represent only one aspect of a complex network of social interactions, and that they are often apparently adapted to the patterns of exchange of resources practised by a society. For example, among Tibetan people the Tre-ba are an élite class who show occasional polyandry in which two or more brothers share a wife as a device to keep their family's wealth together. The transfer of wealth is in fact the key factor determining the pattern of mating in many human cultures. Human mating systems can often be regarded as adaptations to the pattern of wealth distribution in a way analogous to that in which animal mating systems are adapted to the distribution of ecological resources.

We saw in Chapter 3 that in animals the nature of the mating system and the action of sexual selection are inextric-

ably linked. Can we then infer anything from the apparent effects of sexual selection in humans about the nature of our ancestral mating system? Another human universal is that men are slightly heavier, taller and of more sturdy, muscular build than women. Some people have argued, by inference from other species, that this is evidence that we are naturally polygynous, the greater strength of men being an adaptation associated with inter-male competition for the possession of women. A counter-argument to this is that, as we saw in Chapter 3, anatomical differences between the sexes are by no means always the product of sexual selection. In many species, sex differences have an ecological basis and reflect the different ways that the two sexes exploit their environment, for example in the collection of food. It seems clear that for primitive humans division of labour was, and often still is, a very important aspect of reproductive alliances. The woman, with her heavy commitment to a long gestation period, followed by a prolonged period of infant dependence on her for nourishment and protection, is very ill-equipped to face the demands of hunting and gathering food in primitive societies. Her reliance on the man to protect her and provide her with food must have been considerable. The sturdier physique of men is likely to have been much more important in enabling him to fulfil his parental role in a mating alliance than in disputes between men for the possession of women.

Another common feature of human societies, though not a universal one, is that the formation of sexual alliances is controlled by men. In many cultures women are treated as a commodity that men trade between one another. From a historical viewpoint, this is not surprising when one considers first, that, in primitive cultures, men generally control the sources of wealth through their hunting or farming activities, and second, that one of their major considerations is to pass their wealth on to their offspring. Male domination of sexual relationships is also to be expected in view of the potential uncertainty that men have about the paternity of their partner's offspring. It is in the interests of men to restrict the sexual activities of their mates severely. It is significant that in certain cultures where extra-marital intercourse by the wife is an accepted occurrence, the inheritance of a man's property is through his sister, rather than his wife. His sister's children are certain to be genetically related to him while those of his wife may be quite unrelated.

Ensuring certainty of paternity is a very important aspect of social and sexual customs in many human societies. Adultery by women is often punished by very severe penalties, while that by men is often condoned or ignored. Pre-marital chastity of brides is often very important; among the Zulu a girl commands a higher bride-price from her husband if she is a virgin.

Some form of payment by a man or his kin to the relatives of his future wife is a feature of nearly three-quarters of human societies. In no society does the woman's family pay the family of the husband. The practice of paying a dowry involves the passing of wealth to the married couple, not to the husband's family. The involvement of exchanges of money or goods as part of a marriage contract highlights the fact that human mating patterns are bound up with the economic structure of the society of which they are a part. In many human societies male reproductive success is related to material wealth. In those societies which practise polygyny, it is usually those men who hold the largest share of the wealth who have the most wives. There is a clear analogy here between the control of wealth by men in human societies and the control of ecological resources by males in many animals.

An important element in the biologist's approach to studying the evolution of a species' behaviour, is to compare various aspects of the biology of that species with those of closely related species. When we are considering the biology of man, we tend naturally to draw comparisons with our closest living relatives, the great apes. Such a comparison reveals interesting differences between man, gorilla, chimpanzee and orangutan with respect both to their mating systems and to the extent to which they show sexual dimorphism. The males of these four species differ markedly in the degree to which they are larger than the female and in the size of their external genitalia. In the gorilla *Gorilla gorilla*, the male is twice as large as the female, though his genitalia are very small. Gorillas are polygynous and male sexual access to females is achieved by establishing dominance over rivals, a system which has clearly favoured the evolution of large male size. Copulation occurs only occasionally and is initiated by the female, so that the male requires neither large testes to produce or store copious quantities of sperm or prominent genitals for sexual display. The orangutan *Pongo pygmaeus* shows the same anatomical features as the gorilla, though what little is known about their natural sexual behaviour does not suggest that there is much inter-male competition for females. The chimpanzee *Pan troglodytes* lives in large groups in which mating is promiscuous and frequent. Males do not, however, compete for the possession of females, and are only slightly larger than females. Their testes, however, allowing for the difference in body size, are over three times as large as a man's. Their large size is necessary to maintain the plentiful supply of sperm demanded by the high frequency of copulations. The male, who initiates mating, has a large penis which he displays to oestrous females. Female chimpanzees also prominently advertise their sexuality, developing very large perineal swellings when they are in oestrous. It has been suggested that sexuality in the chimpanzee, with its

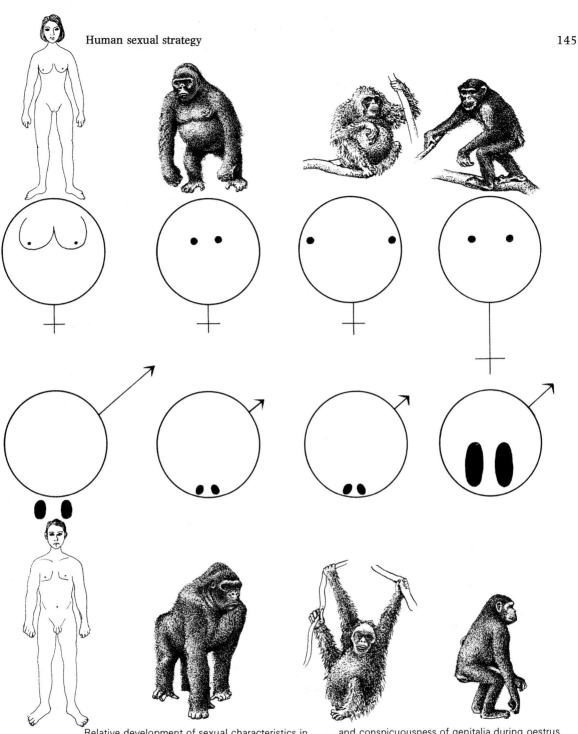

Relative development of sexual characteristics in man and the great apes. The circles are drawn as if all the primates are the same size to show the sizes of sex organs relative to a constant body size. In the female: position and size of mammary glands and conspicuousness of genitalia during oestrus (cross). In the male: size of testes and penis (arrow). The drawings show the relative sizes of the primates. From left, man, gorilla, orangutan and chimpanzee.

elements of promiscuity, overt sexual display, and high coital frequency, has partly evolved as a mechanism that helps to bind the social group together.

In humans, the slightly more powerful physique of men may not, as we have already discussed, have anything to do with sexual competition. Compared with the other great apes, man has much the largest penis. His testes, relative to gross body size, are larger than those of gorillas and orangutans but smaller than those of chimpanzees. The most striking difference between us and our primate relatives is that women develop very pronounced breasts which appear before first conception and which are prominent even when a woman has no infants requiring milk. In virtually all other mammals the mammary glands disappear when they are not required for the feeding of young.

It has been suggested that the large penis of men has evolved as an organ of sexual, and even of aggressive, display. In view of the fact that exposure of the erect human penis is associated, in nearly all human societies, with the privacy that surrounds sexual activity, it seems more plausible that a large penis has evolved to give greater stimulation to the woman. (It must be emphasized here that we are talking about the size of the human penis relative to that of the other great apes. There is no evidence to suggest that variation in penis size among men bears any relation to the intensity of sexual stimulation experienced by women.) An important aspect of human sexuality is that copulation is not an activity that simply has a procreative function, but one which has assumed major hedonistic significance. Humans are highly unusual, possibly unique among animals, inasmuch as they mate at all times, irrespective of whether the female is able or likely to conceive. In the vast majority of species many aspects of sexual physiology and behaviour are adapted to ensure that mating occurs when the female is in peak reproductive condition. There seem to be no such mechanisms in man. One explanation for this is that human sexuality has evolved a secondary function, that of binding the pair-bond together throughout the long period of infant dependence on both the parents. The fact that copulation is pleasurable and rewarding to both partners may serve as a mechanism that reinforces their relationship. It is often suggested that female orgasm is a uniquely human phenomenon which serves to heighten the female's continued interest in intercourse. The fact that men's

testes are smaller, relatively speaking, than those of chimps and larger than those of gorillas suggests that they may be adapted to a pattern of regular intercourse, less frequent than that shown by the promiscuous chimpanzee but more frequent than that of the gorilla for whom male mating success depends on strength.

The fact that the temporal patterning of copulation in humans bears no relation to the woman's oestrous condition may fulfil an additional function. This is related to the fact that women, unlike the females of most other animals, do not signal their oestrous conditions to males. Let us imagine what might happen if they did. Since natural selection would strongly favour those males who devoted their sexual attention only to females who were in oestrus, any female who came into oestrus would become the focus of intense male rivalry. The social disruption that this would create would be severe. In particular it would tend to break up already established pair-bonds, so important for the successful rearing of children. It seems a reasonable assumption that ancestral man was both a very gregarious creature and one who had evolved a pattern of durable sexual alliances. Stable pair-bonds and a cohesive social structure are incompatible if individual females periodically become particularly attractive to males. One evolutionary solution to this problem which, as we saw in the previous chapter, has evolved in the lion, is for females to synchronize their oestrous cycles. It may be significant in this respect that women who live together in student hostels show a tendency to synchronize their menstrual cycles. More important than this is the fact that women are equally attractive and receptive to men at all times and that men receive no overt signals about their oestrous state. In effect, women appear to be in oestrus all the time.

What reproductive strategy is a man to adopt in a situation in which all the women in his group, including his own mate, are sexually attractive and receptive at all times? Since he cannot tell when his own partner is most likely to conceive he is forced to be equally attentive to her at all times, mating with her regularly and maintaining a continual guard against any sexual advances towards her by the other men in the group. The costs that he is likely to incur from any attempt to mate opportunistically with another woman, because of the strong aggression that this would elicit from her mate, are likely to be considerably greater than the benefits that he might gain, in view of the very low and unknown probability that the female would conceive. Thus women, by being sexually receptive at all times and by having cryptic oestrus cycles, reduce the benefits that men might gain by competing to mate with them, and make male fidelity a more adaptive strategy. The fact that the development of women's breasts bears so little relation to their reproductive cycles suggests that breasts have evolved a very important secondary function as stimuli that maintain the sexual interest of the male at all times.

From this rather cursory look at a few aspects of human sexual behaviour it is apparent that there is no simple evolutionary explanation for why we behave the way we do. Some features, such as the universal tendency to form durable pair-bonds and the lack of periodicity in the sexual receptivity of women, can be explained quite plausibly as biological adaptations to the primitive human ecological and social environment. In these respects we can draw direct comparisons between ourselves and other species. However, other aspects of human sexual behaviour, notably the enormous variety of mating systems and customs surrounding sexual alliances, suggest that cultural factors, especially the pattern of wealth accumulation and transfer, are more important than adaptations to our ancestral biological environment.

Here we find ourselves drawing analogies, rather than direct comparisons, between ourselves and other species.

The capacity of the human species to adapt, through technology and culture, to almost any environment on earth, is expressed not only in the variety of human cultures around the world, but also in the changes that have occurred over time within a single society, notably that of western industrial man. It is very apparent that sexual and marital customs and attitudes are changing very rapidly in the west. An increase in divorce rate has brought about a trend away from lifetime pair-bonding towards serial monogamy, and there is a steady increase in the frequency of one-parent families. These changes are the result of many cultural, economic and technological changes, such as the liberalization of sexual attitudes, the increasing economic emancipation of women, and the perfection of techniques of contraception. A particularly important development, which may well have the most profound effects on our social and sexual behaviour, is that the procreative and hedonistic aspects of sex have, potentially, become totally emancipated. As the environment which man has created for himself becomes increasingly removed from that in which he evolved, we may expect that aspects of human sexuality which we may regard as products of natural selection become increasingly less important determinants of our behaviour than economic, technological and cultural aspects of human society.

Glossary

abdomen in adult insects and spiders, the posterior section of the body which lacks appendages; in vertebrates, the part of the body containing the digestive organs. See also **thorax**

algae simple, predominantly aquatic plants; many are single celled and visible only under the microscope, although their effect in turning water green may be visible to the naked eye

altricial totally reliant on parental care or nourishment for an extended period after birth or hatching, *eg* the young of most mammals and perching birds (*cf* **precocial**)

amphibians a class of vertebrates including the frogs, toads, newts and salamanders

anisogamy a type of sexual reproduction in which the gametes are of different sizes, prevalent in the great majority of animal species (*cf* **isogamy**)

antennae in insects, paired appendages of the head which have a sensory function, being sensitive to touch, smell, and in some species, sound

arachnids a class of invertebrates with four pairs of legs that includes spiders, scorpions and mites

cells the units of which living bodies are composed, separated from each other by membranes

cephalopods a class of molluscs characterized by sucker-bearing arms, *eg* octopuses, squids, cuttlefish

chromosomes thread- or rod-like bodies carrying the genetic material and located in the cells of organisms

cleidoic enclosed within a protective shell or membrane, as are the eggs of reptiles, birds and some mammals

cloaca in vertebrates, the common duct of the intestines, bladder and genital organs. Absent in most mammals

convergent evolution the development of similar characteristics in unrelated organisms sharing a common mode of existence

crustaceans	a class of invertebrates which includes the crabs and shrimps
cytoplasm	the material included in a cell, with the exception of the nucleus
divergent evolution	the gradual loss of similarity between related organisms when subjected to different environments
dorsal	of, or pertaining to, the back
embryo	a young animal at a rudimentary stage of development
evolution	a cumulative, inheritable change in a population
family	in biological classification, a group of related genera
fusion	the joining of the nuclei of two gametes to form the first cell (zygote) of the new individual
gametes	the sex-cells of an organism, the sperm and the egg, which contain only half the normal number of chromosomes
gastropods	a class of molluscs characterized by a single, usually coiled, shell and a muscular foot, *eg* periwinkles, whelks, slugs and snails
gene	the smallest, indivisible unit of heredity along the chromosomes
genus (pl. genera)	in biological classification, a group of related or similar species
gestation	in mammals, the period of development of the young in the uterus
hermaphrodite	an organism which has both male and female reproductive organs
hybrid	the offspring of parents from different species or other genetically dissimilar groups *eg* strains, varieties, races
invertebrate	animal lacking a backbone, *eg* worms, insects, crustaceans, molluscs
isogamy	a relatively uncommon kind of sexual reproduction in which the gametes of each sex are similar. This occurs in some protozoans, algae and fungi
mammals	a class of vertebrates that suckle their young with milk and generally have fur
meiosis	the process of cell division by which gametes are formed. In meiosis, the genetic material is rearranged, and the number of chromosomes in each cell is halved
mitosis	the process of cell division by simple splitting of the contents to create two genetically identical daughter cells, each with a full set of chromosomes
molluscs	a group of invertebrates including clams, octopuses and snails; they are predominantly shelled, and with the exception of land snails and slugs, largely aquatic. See also **cephalopods** and **gastropods**
monogamy	reproductive system in which individuals have only one mate in a breeding season

mutation   a sudden change in the genetic material

natural selection   the non-random elimination of individuals (and therefore of genes) from a population

oestrus   part of the reproductive cycle in which the female is sexually receptive

ovo-viviparous   the condition of producing eggs enclosed in membranes which are retained inside the mother, where they eventually hatch. This occurs in many insects and reptiles (*cf* **viviparous**)

parthenogenesis   reproduction by a female without fertilization by a male

pedipalps   in spiders, a pair of head appendages modified for transferring sperm during fertilization

pelagic   living in the open waters of the sea

placenta   a structure in the uterus of mammals where the bloodstreams of mother and offspring come into close contact, allowing nutrients to diffuse from one circulation to the other

polyandry   reproductive system in which the female may mate with more than one male in a breeding season

polygamy   reproductive system in which animals have more than one mate in a breeding season

polygyny   reproductive system in which the male may mate with more than one female in a breeding season

polymorphism   the sustained appearance of two or more, genetically determined, distinct forms within a population

population   a group of organisms of the same species living and breeding together

precocial   birth or hatching in an advanced state of development and capable of living wholly or partly independently of the parent at an early age, *eg* the young of most invertebrates, ground-nesting birds, and some mammals (*cf* **altricial**)

primates   a group of mammals including monkeys, apes and man

protozoans   a very varied group of simple, predominantly aquatic animals in which the body consists of one cell only

reptiles   a class of vertebrates including the lizards, snakes, tortoises and crocodiles

sexual dimorphism   the presence of pronounced physical and behavioural differences between the two sexes of a species

sexual selection   the selective force that arises when certain individuals gain an advantage in mating. This occurs most commonly in males, producing, for example, exaggerated antlers or plumage.

somatic   a term applied to the parts of the body not involved in the production of gametes

species	a group of individuals which show similar features and which can interbreed to produce viable offspring
spermatheca	a sac used for the storage of spermatozoa in the female, particularly in invertebrates
spermatophore	a capsule or sac of spermatozoa passed from the male to the female during mating in many invertebrates, fish and amphibians
stridulation	the production of sound by the rubbing together of parts of the body, common in crickets and grasshoppers
thorax	in insects and spiders, the region between the head and the abdomen (*qv*), bearing the legs, and, in winged insects, the wings
ungulates	a group of widely divergent hoofed animals linked by common adaptations to a grazing existence
urodeles	a group of terrestrial and aquatic amphibians, comprising the newts and salamanders
ventral	the part of the animal normally facing the ground
vertebrate	animal possessing a backbone, *ie* fish, amphibians, reptiles, birds and mammals
viable	capable of living and carrying out normal development and reproduction
viviparous	giving birth to live young that have reached an advanced stage of development while within the mother (*cf* **ovo-viviparous**)
zygote	the first cell of a new individual with a full number of chromosomes, formed by the fusion of male and female gametes

Further reading

Most of the factual material for this book was gathered from original research papers which are too numerous to list here. The most complete account of the topics I have discussed is *Sex, Evolution and Behaviour* by M. Daly and M. Wilson (Duxbury Press 1978). Some aspects of sexual behaviour are discussed in detail in certain chapters in *Behavioural Ecology, an Evolutionary Approach*, edited by J. R. Krebs and N. B. Davies (Blackwells Scientific Publications 1978), particularly those by J. Maynard Smith on the evolution and ecology of sex, G. A. Parker on searching for mates, N. B. Davies on territorial behaviour, S. T. Emlen on cooperative breeding in birds and T. R. Halliday on mate choice and sexual selection. The discussion of mating systems in Chapter 3 is based on a review by S. T. Emlen and L.W. Oring (1977) 'Ecology, Sexual Selection and the Evolution of Mating Systems', *Science* **197**, 215–223, which provides an invaluable starting point for anyone wishing to read more widely on the ecological aspects of sexual behaviour.

Acknowledgements

10 Günter Ziesler; 11 Dr Stevan J. Arnold;
13 Joyce Tuhill; 14 Dr Wayne Aspey;
15 Heather Angel; 17 top and bottom:
Oxford Scientific Films; 18 top and bottom:
Oxford Scientific Films; 22 Masood Quarishy/
Bruce Coleman Ltd; 23 Jonathan Blair/Susan
Griggs Agency; 24 Heather Angel; 25
25 M. P. L. Fogden; 26 Heather Angel;
27 Eric Hosking; 29 Stephen J. Krasemann/
NHPA; 33 Eric Hosking; 36-7 Fransisco
Erize/Bruce Coleman Ltd; 38 Anthony
Maynard; 40-1 M. R. Price/Natural Science
Photos; 42 Heather Angel; 44-5 Leonard Lee
Rue III/Bruce Coleman Ltd; 46 Leonard Lee
Rue III/OSF/Animals Animals; 47 Heather
Angel; 49 H. Rivarola/Bruce Coleman Ltd;
52 Jane Burton/Bruce Coleman Ltd;
54 Heather Angel; 55 Dr J. B. Nelson;
56 Brian Hawkes/NHPA; 57 P. A. Hinchliffe/
Bruce Coleman Ltd; 59 M. P. L. Fogden;
60-1 Dr Nigel Collar; 62 David Hughes/Bruce
Coleman Ltd; 63 Nicholas Hall; 64 Wilf
Taylor/Ardea London; 65 Stephen Dalton/
NHPA; 67 Dr Merlin D. Tuttle; 69 Tony Allan;
70 Philippa Scott; 71 Dr H. Carl Gerhardt;
72-3 Anthony Maynard; 74 Anthony
Maynard after Dr M. J. Littlejohn; 78 Anthony
Maynard; 79 Nicholas Hall after Professor
N. Tinbergen; 81 Jane Burton/Bruce Coleman
Ltd; 83 Eric Hosking; 84 Peter Johnson/
NHPA; 88 James Tallon/NHPA; 89 Hermann
Eisenbeiss/Frank W. Lane; 90 Dr T. Clutton-
Brock; 92 top: P. H. Ward/Natural Science
Photos; bottom: Stephen Dalton/NHPA;
94 Dr N. B. Davies; 96 top: Dr N. B. Davies,
by kind permission of Ballière Tindall;
bottom: Heather Angel; 97 Eric Hosking;
99 Nicholas Hall after Dr Stevan J. Arnold;
101 Leonard Lee Rue III/Bruce Coleman Ltd;
103 Dr Geoff Parker; 104 Oxford Scientific
Films; 106-7 Dr Brian Bertram; 110 Nicholas
Hall after Tim Halliday; 111 Heather Angel;
112 Joyce Tuhill after Dr K. D. Wells; 113 top:
K. G. Preston-Mafham; bottom: K. G.
Preston-Mafham; 115 Joyce Tuhill; 116 top:
Anthony Bannister/NHPA; bottom left:
K. G. Preston-Mafham/NHPA; bottom right:
K. G. Preston-Mafham; 117 Peter Ward/
Bruce Coleman Ltd; 119 Heather Angel;
120 Tim Halliday; 121 Oxford Scientific Films;
122 Heather Angel; 123 Nicholas Hall after
Professor N. Tinbergen; 124 Jane Burton/
Bruce Coleman Ltd; 125 Heather Angel;
126 Dr Stevan J. Arnold; 128 M. P. L.
Fogden; 129 Dr Stevan J. Arnold; 132 M. J.
Coe/Oxford Scientific Films; 135 Philippa
Scott; 139 Dr Brian Bertram; 140 Ardea
London; 141 Alan Hutchison Library;
142 Alan Hutchison Library; 143 Alan
Hutchison Library; 145 Joyce Tuhill after
Dr R. V. Short.

Index